MEDICINE, RELIGION, AND MAGIC
IN EARLY STUART ENGLAND

THE MAGIC IN HISTORY SERIES

FORBIDDEN RITES
A Necromancer's Manual of the Fifteenth Century
Richard Kieckhefer

CONJURING SPIRITS
Texts and Traditions of Medieval Ritual Magic
Edited by Claire Fanger

RITUAL MAGIC
Elizabeth M. Butler

THE FORTUNES OF FAUST
Elizabeth M. Butler

THE BATHHOUSE AT MIDNIGHT
An Historical Survey of Magic and
Divination in Russia
W. F. Ryan

SPIRITUAL AND DEMONIC MAGIC
From Ficino to Campanella
D. P. Walker

ICONS OF POWER
Ritual Practices in Late Antiquity
Naomi Janowitz

BATTLING DEMONS
Witchcraft, Heresy, and Reform in the
Late Middle Ages
Michael D. Bailey

PRAYER, MAGIC, AND THE STARS IN THE
ANCIENT AND LATE ANTIQUE WORLD
*Edited by Scott Noegel, Joel Walker,
and Brannon Wheeler*

BINDING WORDS
Textual Amulets in the Middle Ages
Don C. Skemer

STRANGE REVELATIONS
Magic, Poison, and Sacrilege in
Louis XIV's France
Lynn Wood Mollenauer

UNLOCKED BOOKS
Manuscripts of Learned Magic in the Medieval
Libraries of Central Europe
Benedek Láng

ALCHEMICAL BELIEF
Occultism in the Religious Culture of
Early Modern England
Bruce Janacek

INVOKING ANGELS
Theurgic Ideas and Practices, Thirteenth to
Sixteenth Centuries
Edited by Claire Fanger

THE TRANSFORMATIONS OF MAGIC
Illicit Learned Magic in the Later Middle Ages
and Renaissance
Frank Klaassen

MAGIC IN THE CLOISTER
Pious Motives, Illicit Interests, and Occult
Approaches to the Medieval Universe
Sophie Page

REWRITING MAGIC
An Exegesis of the Visionary Autobiography of a
Fourteenth-Century French Monk
Claire Fanger

MAGIC IN THE MODERN WORLD
Strategies of Repression and Legitimization
Edited by Edward Bever and Randall Styers

The Magic in History series explores the role magic and the occult have played in European culture, religion, science, and politics. Titles in the series bring the resources of cultural, literary, and social history to bear on the history of the magic arts, and they contribute to an understanding of why the theory and practice of magic have elicited fascination at every level of European society. Volumes include both editions of important texts and significant new research in the field.

MEDICINE, RELIGION, AND MAGIC IN EARLY STUART ENGLAND

RICHARD NAPIER'S MEDICAL PRACTICE

OFER HADASS

THE PENNSYLVANIA STATE UNIVERSITY PRESS
UNIVERSITY PARK, PENNSYLVANIA

Frontispiece: Anonymous British artist, *Richard Napier*, ca. 1630. Bequeathed by Elias Ashmole, 1692; WA1898.19. Image © Ashmolean Museum, University of Oxford.

Library of Congress Cataloging-in-Publication Data

Names: Hadass, Ofer, 1969– author.
Title: Medicine, religion, and magic in early Stuart England : Richard Napier's medical practice / Ofer Hadass.
Other titles: Magic in history.
Description: University Park, Pennsylvania : The Pennsylvania State University Press, [2018] | Series: The magic in history series | Includes bibliographical references and index.
Summary: "Explores the work of the astrologer-physician and Anglican rector Richard Napier (1559–1634). Examines Napier's medical and magical practices in their larger context and shows how the physician incorporated both astral and ritual magic into his medicine"—Provided by publisher.
Identifiers: LCCN 2017048628 | ISBN 9780271080185 (cloth : alk. paper)
Subjects: LCSH: Napier, Richard, 1559–1634. | Medicine—England—History—17th century. | Medical astrology—England—History—17th century. | Magic—England—History—17th century. | Theology—England—History—17th century.
Classification: LCC R489.N37 H33 2018 | DDC 610.942—dc23
LC record available at https://lccn.loc.gov/2017048628

Copyright © 2018 Ofer Hadass
All rights reserved
Printed in the United States of America
Published by The Pennsylvania State University Press,
University Park, PA 16802-1003

The Pennsylvania State University Press is a member of the Association of University Presses.

It is the policy of The Pennsylvania State University Press to use acid-free paper. Publications on uncoated stock satisfy the minimum requirements of American National Standard for Information Sciences—Permanence of Paper for Printed Library Material, ANSI Z39.48–1992.

To Inbal, Eviatar, Avishag, and Avigail

לענבל, אביתר, אבישג ואביגיל

CONTENTS

List of Illustrations / ix
List of Symbols / xi
Acknowledgments / xiii
Conventions / xv

Introduction / 1
Chapter 1   Astrological Medicine / 14
Chapter 2   Astral Magic / 60
Chapter 3   Converse with Angels / 86
Chapter 4   Religion and Knowledge / 123
Conclusion / 140

Notes / 145
Bibliography / 185
Index / 209

ILLUSTRATIONS

1. An anonymous sketch of the Old Rectory House in Great Linford / 4
2. Entries in Richard Napier's casebook for June 10 and 11, 1603 / 30
3. Calculations for and description of the making and use of two gold rings with a coral stone / 74
4. Richard Napier's medical notes in the case of John Wyllyson, April 1, 1620 / 78
5. A sheet of Richard Napier's *Interviews* from August 1620 / 90
6. A short disease prognosis attributed to the archangel Raphael / 97
7. Richard Napier's prescription in the case of Sir William Andrews, March 31, 1618 / 100
8. Richard Napier's medical notes in the case of An Foster, April 28, 1613 / 101
9. A page from Elias Ashmole's instructions for summoning angels / 112

SYMBOLS

Astrological Symbols

☉	Sun	♈	Aries
☽	Moon	♉	Taurus
☿	Mercury	♊	Gemini
♀	Venus	♋	Cancer
♂	Mars	♌	Leo
♃	Jupiter	♍	Virgo
♄	Saturn	♎	Libra
☊	Dragon's head	♏	Scorpio
☋	Dragon's tail	♐	Sagittarius
		♑	Capricorn
☌	Conjunction	♒	Aquarius
✶	Sextile	♓	Pisces
▢	Quartile		
△	Trine		
☍	Opposition		

Units of Measurement

ξ   Ounce (or oz)
Ʒ   Drachm (⅛ ounce)
β   Semis (half)

ACKNOWLEDGMENTS

Almost thirty years ago, the combination of an advanced course in quantum physics and Robert Pirsig's popular *Zen and the Art of Motorcycle Maintenance* introduced me to the philosophical quandaries of modern science. While making a career in the field of health information technology, I also embarked upon an intellectual voyage to explore my nebulous doubts. A master's degree in philosophy, guided by Ofer Gal, constituted an initial attempt to evaluate conceptions of truth. It was in one of Gal's seminars that I first became acquainted with Napier and was encouraged to obtain images from his archive. Fascinated by Michael MacDonald's portrayal of a doctor who "lived on the cusp between two eras," I resolved to delve into Napier's career as a litmus test for his period's interplay between magic and science. It is upon the resulting investigation that this book rests.

Three excellent mentors have guided me on this arduous road: Joseph Ziegler, who, with a steady hand and endless patience, steered my academic efforts to their fruitful end; Zur Shalev, who attentively read my many drafts and offered insightful comments and critical suggestions; and Lauren Kassell, who generously shared her expertise as a leading historian of early modern astrology and magic. Employing sharp, instructive criticism, Lauren gave me confidence throughout that my work could make a real contribution to current historical research. I would like to offer special thanks to Margaret Pelling, Nadav Davidovitch, Raz Chen-Moris, Miriam Eliav-Feldon, Naomi Tadmor, Gadi Elgazi, and the late Michael Heyd for lending an ear to my ideas and sharing with me their thoughts. The generous comments of the two anonymous readers of this manuscript are also greatly appreciated.

I was fortunate to share my work in several research workshops for young researchers of early modern history, organized by the Historical Society of Israel and the Zalman Shazar Center. The Graduate Studies Authority at the University of Haifa and the Lea Goldberg Fund at the Hebrew University of Jerusalem gave me generous scholarships that covered some of the expenses of traveling to Oxford and London and of the ordering of images. I am also indebted to the Bodleian Library, Oxford, and to the British Library in London for their accommodating hospitality and for permission to reproduce images from their collections. I will truly miss the long hours of wandering through the folios of Ashmole's weighty volumes, and shorter moments spent cruising the streets of Oxford in brief quests for a bite of lunch. Finally, I am grateful to Robert Ralley and John Young, perhaps the only people in the world who are more versed than me in Napier's handwriting, for their help in figuring out some especially difficult wordings, and to Jay Skork and Simon Cook, who reviewed and refined my English.

My greatest debt is to Inbal, my wife, who coped with many lost weekends and holidays and who constantly encourages me to conquer new peaks. Eviatar, Avishag, and Avigail are still puzzled as to why their father wastes his little free time on such idle pursuits. It is for the four of them that this book was written.

Petahia, April 2017

CONVENTIONS

Quotations from manuscripts and early modern sources are presented as they appear in the original. Early modern spelling changed sometimes, even within the same sentence. I refrained from making any alterations in spelling, retaining the interchange of *i/j* and *v/u* and the *-th* suffix for the past tense. I did, however, expand abbreviations and contractions, including superlinear strokes. The common use of the letter thorn in some manuscripts and books was replaced with the diagraph *th*.

Existing punctuation was usually modernized, mainly by converting full stops to commas. The erratic use of capital letters by Napier and others was retained. Latin phrases or sentences appear in *italics*. A word in doubt is followed by a question mark and is enclosed in square brackets. Other interpolations, including explications of words, also appear in square brackets. Words written above the text were inserted into their place in the sentence. Deleted words were simply omitted.

Biblical quotations are taken from the English Bible in the Authorized King James Version of 1611, unless noted otherwise, and appear with unmodernized spellings.

INTRODUCTION

When Elizabeth Jennings, a girl of thirteen from Isleworth, Middlesex, became "strangely sick" in the winter of 1622 after being frightened by an old woman who allegedly asked her for a pin, bewitchment was immediately suspected.[1] Seventeenth-century witchcraft accusations connected to personal illness were extremely frequent, especially when the symptoms were strange or unexpected, and this poor girl appeared a typical victim.[2] As Elizabeth grew increasingly ill, losing the ability to walk and repeatedly refusing to eat, her parents called for "Doctor Fox,"[3] who took her to London. Often weeping or sighing, she complained of aches in her knees, arms, head, and heart that suddenly moved from one part of her body to another. She became speechless and was often seized with extreme convulsions that caused loss of motion in her right arm. A consilium of learned physicians prescribed her a vomit, a letting of blood, and a bath of oil; yet such treatment wrought no amendment. Suspicions of witchery were confirmed when, during one of her trances, Elizabeth announced that she had been forespoken by four women, chiefly one identified as Margaret "Countess" Russell:[4] "Jane Flower, Katharin Stubbs, Countesse, Nan wood, These have bewiched all my mothers children. East, west, north and south all these lye. All these are damnable witches.... Put Countesse in prison this child wilbee well. If she had bin long agoe all th'other had bin aliue. Them she bewiched by a catsticke. Till then I shall lye in great paine. Till then by fitts I shalbe in great extremitie."[5] Following this accusation, "Countess" was arrested and imprisoned, and a formal court inquest was initiated. Yet despite some short-term improvement, Elizabeth's bouts with hysteria continued well into the summer, and Lady Jennings, desperate to cure her daughter's condition, called on the help of Richard Napier, a medical practitioner from Great Linford in Buckinghamshire.[6]

A self-made astrologer-physician, who was by then known across the Southeast Midlands as a successful healer and a master of occult arts, Napier was treating around thirty patients a week. Although he was accustomed to cases of suspected witchery, as well as to clients from the ranks of the gentry, it was nonetheless unusual for him to see a patient after unsuccessful treatment by members of the College of Physicians of London. "Hearing that I have helpen many in that kind," he remarked rather complacently in his medical diary, Lady Jennings "hath written to me to come to London."[7] Taking appropriate extra care, Napier documented at relative length noteworthy events from Elizabeth's recent medical history and drew an astrological chart for each of these significant occasions, as well as for the child's nativity. Beside each astrological figure he noted important planetary aspects and made some astrological calculations that showed that Elizabeth was suffering from *epilepttia matricis* or *morbus matricis*.[8] Such astrological judgments stood at the core of almost every medical encounter in Napier's clinic, although in more typical cases a single chart was cast for the moment of the question. As in other astrological-medical practices, Napier used astrology as a tool to reveal the inherent cause of disease, determine an appropriate course of cure, and assign prognosis. It was an instrument of his practice, embedded within the more traditional view of disease as an "imbalance" or "corruption" of bodily humors.

Napier's exceptional reputation, however, was mainly due to his alleged ability to consult with angels, especially the archangel Raphael.[9] Elizabeth's case demanded such extraordinary means, and Napier confronted Raphael with a query that is now bound toward the end of the same medical diary: "Wheather young ms Elis[abeth] Jennys of 13 y[ears] be bewitched or only be troubled with the epilepsye of the mother & [whether] I shall cure her & howe & wheather on[e] called goody countis be a wytch and hath done it." He quoted Raphael's harsh reply—"[she] will dye halfe a yere henc"—alongside the question and copied it word for word into the original case record.[10] Three days later, probably following another spiritual conference, he noted a similar judgment, this time confirming that the girl was indeed bewitched. Underneath this more elaborate response, however, appears an apparently retrospective remark, announcing that the allegations "proved all false."[11] The source of this surprising contradictory judgment is not stated. As in many other cases, angelic guidance provided Napier with a sense of confidence about the facts of the medical situation, gained through the trustworthy testimony of a higher member of the cosmic order. In this particular episode, however, it seems that angelic authority failed to assist him in clearing the allegations of witchery.[12]

Elizabeth lived beyond the six months granted her by the angels, and her mother continued to consult Napier, making further use of his versatile repertoire of treatments to promote her children's health. In the autumn of 1631 Napier prescribed Elizabeth an astrological talisman ("sigil") engraved with the characters of Jupiter. The same magical cure was prepared for her brother Thomas, who "vppon mistaking put his [sigil] on a great while before the houre."[13] Astrological talismans were a form of natural magic: part of Napier's arsenal of magical and spiritual cures that included the use of sigils, pomanders, charms, prayers, and exorcisms. Their employment was based upon the conviction that astral influence operated through the same principles of sympathy and antipathy that were utilized in traditional learned medicine, and that certain astrological symbols, when cast at select times on the appropriate metal or stone, could be used to draw down celestial forces to effectively combat disease.

An eclectic approach such as Napier exhibited in the Jennings case was not an early modern novelty. Ancient and medieval medical practitioners had also offered a combination of natural treatments together with magico-religious ones. However, the profusion of medical theories that characterized the sixteenth and seventeenth centuries brought such eclectic practices to a peak.[14] The expanding use of astrology and alchemy in medicine, the increasing utilization of chemical remedies, and the revitalization of certain magical and "hermetic" traditions significantly widened the range of knowledge systems and healing methods available to the early modern physician. Napier's position as the Anglican rector of his parish, purportedly obliged to assume a more orthodox outlook and abstain from unsanctioned occultism, only emphasizes the puzzle inherent in his recourse to such an amalgam of partly suspect techniques.

Inspired by the practical and intellectual history of an era characterized by a conceptual change of attitudes, I have set out to examine how Napier's seemingly eclectic reconciliation of such diverse approaches was possible. Given the increasingly fragmentary narrative of early modern English history, a broad investigation into Napier's life and work seems to challenge current approaches to this question. Writing from the perspective of the history of medicine, I employ Napier's engagement in certain intellectual and practical traditions in order to suggest that his versatile endeavor was in fact guided by a perfectly coherent view of nature and God. Nonetheless, his creative use of assorted practical means contributed to his immense popularity, attracting thousands of patients to the pastoral meadows of North Buckinghamshire.

Only a few stone houses of the seventeenth-century village remain, contained on either side of High Street in Dovecote Croft, Great Linford. Set back from this quiet lane stands the most significant remnant of the village's glory days: an impressive stone building known as the "Old Rectory." Much altered since it housed Napier's residence and medical practice four hundred years ago, it has an old tile roof, a central oak door, and four ornamental brick chimney stacks. Going round the big house, which is now vacant, one can only imagine the noisy bustle of the seventeenth-century medical practice, crowded with worried patients and housing overnight guests and long-term apprentices. Heading north from the Rectory, through the old manor gates, a yellowish track crosses a large meadow toward the renovated Manor House, which once belonged to the Napiers.[15] A footpath and a two-minute walk uphill lead to the parish church of St. Andrew's (fig. 1). Surrounded by fragrant flowers and ancient graves, the main building of the church almost hidden among tall maple trees, its thirteenth-century tower seems to rule over the entire area. In the recently decorated interior, Napier is named last in an improvised list of medieval and early modern rectors hung on the wall near the altar.[16] This short walk, from his residence and medical practice in the Old Rectory to St. Andrew's Church, where he served as the local priest, provides a good parallel to Napier's life story, linking his separate, yet inextricably related vocations of religion and medicine.

Born in Exeter in 1559 to Ann (Agnes) and Alexander Napier (alias Sandy), the younger son of a noble Scottish family from Merchiston that came to England in the reign of Henry VIII, Richard Napier had four sisters and was the youngest of three brothers.[17] Having no children of his own, he kept in close touch with most of his siblings and nephews, particularly with the family of his elder brother Robert, whose second son, Sir Richard Napier, was to become

Fig. 1 An anonymous sketch of the Old Rectory House in Great Linford, on a stone-mounted information board just outside St. Andrew's Church, produced by the Milton Keynes Parks Trust and Milton Keynes Archeological Unit, Buckinghamshire County Council, August 1992.

Napier's protégé and inherited his entire property and medical practice.[18] Not much is known about our Napier before he arrived at Great Linford, but in a rare account of his life's accidents he recalls how, when he was a boy, his mother, "casting a candlestick in her anger to one of my sisters, . . . steeping [sic] the candle stick lighted vppon my head being a sleepe in my bed, & I blead sevrely vppon it."[19] Napier entered the University of Oxford at the age of eighteen to study theology, receiving his BA in 1584 and an MA a couple of years later. He was elected a fellow of Exeter College in 1580, a position he held until he moved to Great Linford.[20] "When I was a scholer in Oxford," he proclaims in the same autobiographical note, "[I was] mutch given to study, no mynd to marriage, very earnest in disput[e]."[21]

Admitted to the parsonage of Great Linford in May 1589 with the help of some family connections, Napier substituted the challenging academic setting of the university for the seemingly calmer position of a country vicar.[22] Yet "extreme fearfull by nature to declayre or preache,"[23] as he himself admitted, he soon hired a curate to handle some of his ministerial duties for him. Napier first became engaged in the practice of medicine in 1596, after seeking the aid of the famed (yet notorious) astrologer-physician Simon Forman in some medical and personal matters.[24] Extending what might have been an existing interest into an actual occupation, he spent several periods at Forman's house in Lambeth, London, learning the secrets of astrological medicine and geomancy.[25] Back in Great Linford, Napier started to resolve medical questions for his relatives, acquaintances, and neighbors, rapidly expanding the scope of his practice into a full-fledged vocation. Treating both commoners and gentry from an area much larger than that of a modern general practitioner, he established a prosperous career in medicine that lasted until his death in 1634.[26] Thanks to his influential patron, Lord Thomas Wentworth, and as a physician practicing in rural Buckinghamshire, Napier managed to sidestep the medical politics of London and avoid confrontation with its institutional authorities.[27] Consulted by fellow practitioners and hosting aspiring apprentices who came to Great Linford to learn the secrets of medicine and magic, he became known over time as a most skillful physician and adept in occult arts.

Napier systematically recorded the consultations he held with his patients, leaving behind one of the most outstanding recorded examples of early modern medicine and certainly the richest one known. This huge collection of medical records—almost sixty volumes of documented medical encounters—was left upon Napier's death to his nephew, Sir Richard, and, after the latter's demise in January 1676, was handed by his son Thomas to Elias Ashmole, a renowned antiquary and charter member of the Royal Society, who bound, numbered, and organized the papers.[28] Together with the rest of Ashmole's enormous collection of occult material, Napier's medical diaries were housed in the Ashmolean Museum, Oxford, until 1860, when they were moved to their current dwelling in the Bodleian Library. The generous gift was not always appreciated, as is evident from a letter written in 1693 by the second keeper of the Ashmolean Museum to one of its benefactors: "You take care to send us nothing but what is valuable & pertinent. But I could heartily wish Mr Ashmole had also done the same in his Legacy of Books; & instead of many MS volumes of Mr Napiers Astrological Practice in Physic, & above five hundred other Astrological books, I wish he had given us 50 of his best books relating to coyns and other antiquities, & to Natural Philosophy. Tho' his donation be in its kind, also very usefull & considerable."[29] A minority of the medical documents (mostly explicitly magical material) was probably kept by Napier's grandnephew Thomas; these documents later found their way to the Sloane Collection in the British Library in London.[30]

The contribution of Napier's archive to the body of evidence on early modern intellectual life also includes numerous alchemical papers, correspondence on astrological, theological, and personal matters, a few unpublished pamphlets and expositions (mostly theological), prayers and invocations, drafts of sermons, scattered autobiographical notes, and miscellaneous memoranda, especially on the lending of money and books.[31] Constantly engaged in ancient and contemporary learning, Napier was continually borrowing, copying, and reading an endless list of books and manuscripts. His surviving letters and personal notes reveal his close relations with members of his extended family, nearby gentry, and professional colleagues. Napier's conversations with angels, traced in the curious, prophetic remarks and clusters of queries that populate some of the diaries, compose another intriguing facet of this versatile collection. Focused on a single person, this unique assemblage of medical diaries, letters, and other papers thus holds evidence for a three-dimensional discourse: physician-patient, professional, and spiritual.

Despite its accessibility for research, this vast treasure trove of a documented seventeenth-century astrological and medical practice has received limited academic attention. In his landmark book *Mystical Bedlam: Madness, Anxiety, and Healing in Seventeenth-Century England* (1981), Michael MacDonald provided a thorough analysis of some of Napier's patients, diagnoses, and cures, but focused almost entirely on his mentally disturbed patients: a group that comprised less than 5 percent of his clientele.[32] Meticulously breaking his 2,500 cases of mental disorder into groups by age, gender, geographical distribution, social status, symptoms, and diagnoses, and situating them in their relevant social contexts, MacDonald formulated a new understanding of troubled minds in premodern England. Quite unjustifiably, however, this has established Napier's reputation as a "healer of mind."[33] MacDonald concluded that seventeenth-century healing represented a more sympathetic attitude toward mental afflictions than was later permitted when the church and the social elite made insanity a social and medical problem. Ronald Sawyer's unpublished PhD thesis, submitted in 1986 under the supervision of MacDonald, dealt with the broader aspects of Napier's work, providing a wide review of healers, disease, and medicine in this particular setting. Sawyer found that patients demonstrated great flexibility in moving among the various kinds of healers (in what he termed "the patient shuttle") and in utilizing diverse cosmologies in their search for a cure.[34]

When MacDonald wrote his book in the early 1980s, a postmodernist preoccupation with social constructivism prompted the viewing of early modern beliefs and practices as carrying their own intellectual coherence. This shifted the focus in the history of medicine to social forces such as religion, politics, and power.[35] Some social historians set out to analyze case records and manuscripts of early modern practitioners and to examine the context in which they operated, evoking suggestions that "patient records may well prove to have their greatest utility in permitting a systematic exploration of the relationship between medical ideas and medical activities."[36] Others, like Sawyer, went on to highlight patients' perspectives of disease and treatment.[37]

The landscape of early modern medical care has also been widely researched and mapped. Margaret Pelling, Charles Webster, and others have documented the variety of medical practitioners at work in early modern England, showing that medical expertise was very widespread across society.[38] It is recognized that learned and authorized physicians constituted only a small fraction of medical practitioners and that large cities attracted a wide diversity of healers: empirics, barber-surgeons, apothecaries, herbalists, cunning men, uroscopists, and midwives. Even smaller towns and villages were usually inhabited by folk-healers and wisewomen who provided medical care.[39] It was also not unusual for priests to be physicians as well,[40] provoking criticism of clergymen who neglected the soul for the sake of curing the body.[41] Exploring the competitive medical arena of seventeenth-century England, certain scholars employed the model of the "medical marketplace," which allowed the different groups of practitioners to be viewed on an equal footing.[42] More recent studies have stressed social and cultural forces in the practice of medicine and the ways in which practitioners and patients perceived disease and healing and negotiated among themselves.[43]

At the turn of the twenty-first century, historians were still complaining that this important work in the social history of medicine had nevertheless neglected the central aspects of medical knowledge and practice: the actual treatments, explanations, and daily advice that patients were given by their healers.[44] This gap has been partly filled during the last two decades, in works that have analyzed the minutiae of contemporary practice. Focusing on individual case studies instead of grand narratives, these studies examined the making of practical knowledge, its modes of implementation, and the growth of medical authority and reputation.[45] This produced a view of early modern historical figures as active, flexible agents, operating within a changing and vibrant environment that demanded continuous adaptations to competing social, professional, and practical requirements. At the same time, it resulted in an increasingly fragmentary rendering of seventeenth-century medicine and of individual intellectual outlooks.

Although practice is at the heart of much current research, a biased focus on medical ideas, voiced theoretical stands, and written guidelines fuels, I believe, this more piecemeal account. The ascendancy of the written word, especially for historical research, still veils the facts of everyday practice and its underlying ideologies. An ample investigation into Napier's vast archive of practical astrological medicine thus frames a better understanding not only of early modern medical practice, but also of contemporary thought. The vast abundance of carefully documented medical records, the symbols and signs of astrology and astral magic, and the scattered traces of Napier's spiritual interviews were my clues in sketching an intellectual biography for this ambitious and informed practitioner. While his lack of digested works constitutes a challenge to any attempt to make sense of these practical methods in a broader context, it offers nonetheless an exceptional opportunity to move from the particulars of practice to the generalization of a theoretical framework.

Astrological outlooks influenced and sometimes ruled the most forward-looking sectors of early modern culture. A publicly practiced art, astrology played an important part in the arts and science curricula of major universities and informed every aspect of the lives and work of many educated men.[46] It was present in the most fundamental aspects of Napier's work, from his techniques of diagnosis, prognosis, and treatment through the structure and content of consultations to his meticulous record-keeping. The foundations of astrological medicine rested upon the interaction between celestial influences and humoral physiology, the cornerstone of Galenic medicine. Together with other types of astrology, astrological medicine flourished during the first half of the seventeenth century in a burst of publications, studies, and practices, only to decline in legitimacy toward the century's end.[47] Because medical astrologers were mostly visited by their clients (in contrast to classically trained physicians, who usually attended their patients), most astrologer-physicians treated significantly more clients than did other medical practitioners, and the stress astrology laid on time, identity, and age prompted them to make systematic records. In an age when experience was a critical factor in medical practice, this gave astrologer-physicians an important advantage over other contemporary healers.[48]

A succession of scholars throughout the second half of the twentieth century identified "occult philosophies," including astrology, alchemy, and magic, as legitimate research subjects in the context of early modern science, showing how they remained vital ingredients of advanced thought at least to the end of the seventeenth century. This notion, initiated by Lynn Thorndike's description of medieval alchemy and magic as the ancestors of experimental science, was elaborated by Frances Yates, who, in what came to be known as the "Yates Thesis," argued the same for "hermetic magic" and shattered traditional boundaries between regression and progress, superstition and science.[49] Nevertheless, Yates still held strong views on the seventeenth century as a watershed between a shady magical past and modern, enlightened scientific thought. While most historians of science rejected the majority of Yates's arguments, especially the notion of "hermeticism," which came to be seen as a vague and indistinct term, they nonetheless acknowledged "occultism" as part of early modern natural inquiry and as a legitimate area of study for the history of science.[50] The idea of discontinuity and the periodization of early modern history was also questioned, and a new emphasis was placed on the apparent similarities between the theories and practices that ruled both ends of the period.[51]

In his seminal *Religion and the Decline of Magic*, Keith Thomas made the first comprehensive attempt to study occult practices, not as superstitious survivals of a primitive age, but as valid intellectual schemes that were once widely accepted.[52] Considerable academic attention has since been given to the study of magical traditions and texts and their intellectual affinity with the emerging new science.[53] Much less emphasis has been afforded, however, to the actual methods, rituals, and practical considerations of contemporary magi, particularly those involved in talismanic practices.[54] Angelic conversations, of the kind conducted by Napier, have also captivated the minds of modern scholars. Yet an almost exclusive focus on the diaries of John Dee, the narrator of a new religion, has only recently been superseded by the foregrounding of angel conjurers with a more practical agenda.[55]

Appraisal of the role of Reformed Christian thought in the early modern alteration of attitudes, and particularly in the so-called disenchantment of the world, has also undergone a generational historiographical shift.[56] Four decades ago, Thomas argued that the Reformation played a critical role in the rejection of all sorts of supernatural enchantment (specifically in their ecclesiastical manifestations), facilitating the transition to a more "rational" view.[57]

Pointing to evidence of continuities between medieval and Protestant mentalities, Robert Scribner emphasizes, however, the limits of disenchantment in the sixteenth and seventeenth centuries and argues that the Reformation did not bring such a dramatic break with the Catholic past. Rather, he suggests, it underlined the role of nature in retrieving divine messages to humanity while producing, at the same time, religious and magical rituals of its own.[58] More generally, a new understanding of the prominence and innovativeness of Renaissance magic has seriously called into question the notion of a linear process toward disenchantment and the use of reason. Instead, Alexandra Walsham suggests, we should be "thinking in terms of cycles of desacralization and resacralization, disenchantment and re-enchantment," that last even to this day.[59]

An Anglican priest and a learned scholar in theology, whose scant theoretical writings deal almost exclusively with religious issues, Napier provides through his theological views the necessary glimpse into his broader outlook.[60] Reconstructing and situating his ideas within contemporary religious atmospheres thus constitute a vital step toward understanding his approach to the exploration of nature and God. At the turn of the seventeenth century, the famous Elizabethan Settlement was already firmly established in English religious culture. Yet the celebrated "Golden Age" of the Elizabethan era was in fact a time of bitter struggle over fundamental questions of belief, worship, and ecclesiastical polity that would only intensify in the reigns of the Stuarts. Jacobean England was predominantly Protestant, albeit in its own unique way, but religious undercurrents and ideological conflicts were still very fierce. Proponents of further reformation, mostly guided by the Genevan Calvinist model, demanded a complete abolition of Catholic practices and the rearrangement of ecclesiastical structures, while their opponents defended the traditions and rituals that were left intact by the first round of reforms or even lobbied for the return of some discarded doctrines. The fortunes of these conflicting powers changed several times over the first decades of the seventeenth century, and opinions that were fiercely criticized by the ecclesiastical leadership in the early 1600s could be regarded as orthodox in the 1620s and vice versa.[61]

Just as with the emergence of the new sciences and the impact of the Reformation on Western Christian thought, so too the study of the post-Reformation English church has seen a comparable turn from inclusive historical accounts to more local, discrete narratives. The Whig and Marxist interpretations, which portrayed enlightened "Puritanism" as a distinctive force of historical change—either making way for the rise of capitalism or as a bourgeois revolution against the repressive Stuart church[62]—gave way in the 1960s and 1970s to more dynamic readings of the theological conflicts that occurred over the course of the seventeenth century.[63] The "Anglican vs. Puritan" opposition still dominated understanding of post-Reformation English Christianity, but its clear-cut distinctions were gradually beginning to blur. The apparent ease of movement across boundaries, from conformity to nonconformity and back to conformity, supported the suggestion that contemporary religious behavior might be more properly described in terms of outlook and style of piety than in the fixed and constant terms of a movement, sect, or set of beliefs.[64] "Puritanism," to name one particular strand of religious attitudes, "could therefore appear and reappear like the smile on the face of the Cheshire cat, according to the circumstances of local and national politics."[65]

As the discussion above demonstrates, grand narratives concerning the intellectual and cultural change that allegedly took place in the early modern period have given way to a series of localized, disparate stories. In the case of medicine in particular, scholars now tend to highlight the broad range of ideas and practices and the simultaneous reference to competing philosophical and scientific theories as evidence for inconsistency of thought. An astrologer-physician, alchemist, Anglican clergyman, and conjurer of angels, Napier seems to stand out as a strikingly genuine example of such excessive eclecticism. Like most educated physicians, he viewed the body in terms of a Hippocratic-Galenic system of flowing and intermixing humors, providing his patients with traditional herbal cures that provoked their evacuation. At the same time, he also utilized alchemical and mineral cures that he had bought or prepared by himself, employed astrological sigils and aromatic pomanders, and often had recourse to prayer.[66] He gained medical insight by means of astrological judgment, but also used urine analysis, spiritual conferences with angels, geomancy, aeromancy (interpreting atmospheric conditions), and arithmancy (analyzing the numerical value of a person's name).[67]

Historians have offered differing interpretations of this apparently eclectic approach, reflecting in a sense some of the broader historiographical trends I have traced above. MacDonald suggests that early modern medical practices such as Napier's "mirror the pragmatism of English physicians . . . who were chiefly concerned to discover effective medical treatments regardless of their sources."[68] Sawyer, shifting the focus to the patient's gaze, argues for a flexible and multicausal cosmology in the early modern world, maintaining that patients, as well as practitioners, were willing to adopt different explanations and views about disease and treatment according to the situation at hand and as a part of their efforts to find a cure.[69] Andrew Wear maintains that the sense of continuity in practical medicine was stronger than the apparent conflict between rival ideas. Some opportunistic empiricists, he argues, simply realized that their clients would respond favorably to practical methods that could meet their own expectations or even to the reference of multiple intellectual schemes.[70] A more recent and probably bolder approach suggests that the distinctive mentalities that characterized the different early modern traditions of thought allowed those who stood in theoretical conflict to "compartmentalize their minds" in order to employ the separate concepts and methods of various domains at different points of time or in varied contexts.[71] Acknowledging the legitimate place of ambivalence and contradiction in both theory and practice, Alexandra Walsham draws attention to "the general untidiness of the human mind . . . [and] the eclectic intermingling of inconsistent opinions that is a perennial feature of individual and collective mentalities."[72]

The image of early modern medicine has never been more fragmentary, forcing historians to look at microhistories, not only at the individual level, but even in relation to specific projects, contexts, or time frames. More than ever, historical research is in need of a reconstruction, if not of the intellectual landscape of medicine itself, then at least of some of the "occult philosophies" that seem to have governed large portions of its contemporary application. Through the three distinct lines of inquiry that I portrayed above—medicine, magic, and religion—I will draft Napier's practical and intellectual biography within this broader context. Where other historians have emphasized medical theories, I have identified what amounts to a thesaurus of practical knowledge that drew upon the amassed contributions of ancients and moderns and Napier's own experientially derived wisdom.

Opening with the structure and content of his astrological-medical consultations, I consider Napier's conception of body and disease, reconstruct the charting of astrological figures, and examine their role in his methods of diagnosis, treatment, and follow-up. Medical astrology, I argue, supplied Napier with a simple, fast, and reliable way of knowing, which easily outperformed the complex and ambiguous method of symptoms and signs on the one hand and the superficial symptomatic medicine on the other. Representing a specific strand of early modern medicine, it created a parsimonious and uniform nosology within the context of humoral medicine, supplying Napier with an effective method for reading a narrative of disease that paralleled that of contemporary patients.

Turning to his more "esoteric" practices of astral magic and interviews with angels, the second and third chapters deal with Napier's involvement in certain magical traditions. Exploring the practical considerations that guided his employment of astrological talismans, I portray this powerful technology of magic as part of his regular medical practice. Napier's spiritual interviews, carried out mostly with the archangel Raphael, afforded him a further source of certified knowledge on the troubles and fortunes of his distressed clientele.

I suggest that these conversations were spiritual self-experiences much more than they were encounters with supernatural beings and that, set within a wider astrological outlook, they were perfectly consistent with Napier's philosophical worldview. Both kinds of magic assisted Napier in gaining a better understanding of the cosmos and in utilizing its forces for the sake of medical care.

The last chapter shifts focus to Napier's theological outlook, using his handful of letters and tracts as a key to deciphering his corresponding views on religion and nature. Placing him within the context of contemporary religious conflicts, I show how Napier's style of piety adjoined traditional and Reformed modes of thought to a Christian-Neoplatonic outlook. I also argue for the significance of his quest for infallible knowledge, drawing on the cumulative efforts of past generations and the "general consent" of learned men. Allowing ancient wisdom and elements of natural philosophy into the Christian narrative, yet armed with the spirit of active inquiry, Napier provided an effectual contribution to the contemporary venture of putting existing knowledge to useful ends. Gazing at a world filled in every corner with elements of divine revelation and godly hints, he adopted a harmonious view of creation that fostered the construction of a state-of-the-art seventeenth-century medical practice.

1

ASTROLOGICAL MEDICINE

Discussing the case of "a sick doctor" in his celebrated *Christian Astrology*, William Lilly erected an astrological figure, used it to identify the afflicted parts of the body, and explained the cause of the illness as well as the malignant constitution of humors, only to conclude, "Having my self little judgment in Physick, I advised him to prescribe for himself such Physicall Medicines as were gently hot, moyst and cordial, whereby he might for a while prolong his life; for the ☽ in the fourth in ✶ with ♄, argued sickness until death."[1] Astrology and medicine, two arts that intersected in ancient times for practical purposes, achieved a peak of collaboration in the work of many medical practitioners of the sixteenth and seventeenth centuries. Although both diffused vital components into this amalgamation, each still maintained its own individual role. In the above quotation, Lilly draws the line between the two practices in the boundary between knowing and acting, that is, between diagnosis and prognosis, on the one hand, and prescribing a treatment, on the other. As Napier held probably one of the busiest practices of astrological medicine in early modern England, a close investigation into his medical transcripts offers a unique opportunity to reconstruct the practical facets of this almost forgotten art.

Over the last thirty years the history of medicine has increasingly focused on medical practice and on practitioners.[2] Yet unveiling what clinicians actually did does not necessarily disclose meaning or reveal the theories that guided them. To decipher the meaning of their actions from methods of behavior based on actual medical records, these records must be interpreted in the light of information collected from other discursive sources.[3] This is particularly true in Napier's case, as reconstructing his practical methods without the benefit of any explanatory texts by him requires the consideration of other didactic sources. Hence, although my analysis relies heavily on Napier's case records, it also rests on the astrological and medical guides of Simon Forman, Napier's astrological teacher, and on several other manuals of astrological medicine from the sixteenth and seventeenth centuries.[4] The sample of cases that underlies my investigation in this chapter is primarily (but not solely) based on the medical consultations Napier performed during the years 1602 and 1603, which are, however, representative of his entire career.[5] At this time he was already beginning to establish his own unique method of treatment, which involved some departure from his tutor. Nevertheless, he was still very much influenced by Forman's guidelines, and it therefore makes sense to rely on Forman's manuals when studying Napier's medical records. Most of these records are fairly schematic and have a rather organized structure: patient's name, place of residence and age, date and time, an astrological figure, a list of complaints or medical problems, and prescribed remedies.[6] Through a structured examination of these notes I have traced the theoretical and methodological foundations of Napier's medical practice, which I present in this chapter.

I will open with a portrayal of Napier's theoretical framework: his views on the human body and disease, and the basic concepts behind his practice of astrological medicine. I will then examine the individual components of his medical consultations in the same order as they usually occurred in the course of his work: from recording patient and encounter details through collecting evidence to assessment, judgment, and treatment. In all these steps, medicine and astrology were inextricably linked.[7] The significance of astrology in Napier's medical practice lay primarily in its ability to provide a quick and unambiguous path to medical judgment. Put in terms of the distinction between knowing and acting, revealed in Lilly's quotation above, astrology supplied Napier with the necessary practical knowledge within the medical encounter. Yet Napier's astrology and astrological worldview were not merely functional tools in his consulting room. As we shall see, they offer a first key to unlocking Napier's philosophical and theological outlook.

The Theoretical Framework

The Image of Body

Shortly after noon, on one of the first days of 1603, a forty-year-old woman named Mary Barly [Barlee], from the small town of Woburn, arrived at Napier's consulting room in Great Linford. Not having had her period for over two months and suffering from aches, swellings, and a sense of vertigo, she hoped to find relief with the help of the already renowned doctor. After noting her name, place of residence and age, and the exact date and time of her visit, Napier recorded in his medical diary: "[M]enses obstructi since alhallowtide. they ly about her hart & stomacke. a swelling in her leggs & hands. an ache, lightnes and swimming in her head. once a month a drop of bloud at [the] nose. Alhand[al] Jeralog [hiera logadii] ana ʒβ [of each half a drachm], decoct[ion] Puleg[ium] ana ξiβ [of each one ounce and a half], xij d[imes]."[8] This brief medical note provides a glimpse into Napier's view of the human body as an arena of inner and outer flows, always in danger of obstruction, overflow, or loss of way. His report begins with a portrayal of the patient's menstrual disorder and its exact duration. What follows can be understood as referring to the place of the "lost" menses, or more generally to the location of the onset of the disease. The closing observation, concerning the monthly nosebleed, points to the strange path the menstrual blood took into the woman's nose, probably in its attempt to find an orifice through which to escape.[9] As I will show in the following section, this is a typical example of how flows and obstructions were used to explain healthy and diseased functions within the early modern body.

In order to categorize the parts and functions of the human body, modern medicine uses the concept of "body systems" (digestive, cardiovascular, respiratory, and so forth), all of which work together to constitute the living human body. This specific categorization implies a set of functionalities, each resting on a group of organs that have specific functions in its operation. The link is unbreakable: the respiratory system cannot operate without the lungs, and the lungs are one of the building blocks of the respiratory system, together with the respiratory muscles and the diaphragm. By contrast, early modern physicians, basing their understanding of the body mainly on Hippocratic texts from the fourth century B.C., gave a global and primary role to four bodily fluids called humors (blood, phlegm, melancholy, and choler). These humors corresponded to even more basic aspects of nature: the four Aristotelian elements (air, water, earth, and fire), the four combinations of primary qualities (hot and moist, cold and moist, cold and dry, hot and dry), the four seasons of the year, and so on. Hippocratic humors were not part of any specific body system, but rather constituted an autonomous layer in the image of the body. Galen (second century A.D.), the most influential physician of later antiquity, incorporated the humoral body view within a much wider anatomical-physiological framework that was based on numerous animal dissections he performed. His view stressed Aristotle's concept of the body as a set of interconnected organs over the Hippocratic view of free-flowing fluids.[10] Yet rather than challenging the Hippocratic paradigm, Galen's variation appended a new layer to the existing image of the human body and its inner processes. Hence the complexity of the ancient and later early modern view of the body, in which recognizing the anatomical structure of body organs did not negate its perception as a container of inner flows. The ambiguity is immediately apparent in Napier's medical records, where affected body parts are listed for almost every case, while, as I shall show later, medical judgment and prescription mainly relied on the state of bodily fluids. This double-sided approach was shared by many of Napier's contemporaries and was echoed in numerous medical handbooks that discussed problems from head to foot while still maintaining the concept of humoral activity and its overall impact on the body.[11] In her influential study of Dr. Johannes Storch, an eighteenth-century German physician from Eisenach, Barbara Duden revealed that the view of the interior body as "an unstructured osmotic space" was still shared by physicians and laypeople as late as the eighteenth century.[12] Her work has been reinforced by several other studies that portray the early modern body as an "interconnected whole," in which bodily fluids and vapors (produced by the flowing humors) traveled around quite freely.[13] Of course, postmortem dissections of human bodies revealed a static anatomical structure, in which the distant body parts were connected only through certain pathways, but this was not seen as limiting the possible functionalities of the living body.[14] Despite rigorous and systematic anatomical research conducted from the fifteenth century (with early beginnings in the thirteenth century), this discrepancy between anatomical findings and the received view of the living body still existed throughout the early modern period.

Traditional research on humoral theory has mainly considered the balance between the four basic humors and the influence of their imbalance on human health. More recent studies, however, have revealed a much more colorful image of bodily fluids. Phlegm, choler, melancholy, and blood, it is now acknowledged, took different shapes in a variety of transformations and forms, were frequently mixed with each other, and could be abundant or lacking in a specific location in the body.[15] Moreover, any humor could be transformed into another, or change to become putrid, venomous, corrupted, or corroding, or induce sickening and dangerous fluxes that had to be evacuated. It was pointed out that even Galen had recognized many possible transformations and variations of the humors and declared that while sometimes humors flow unmixed, at other times they could transform from one to another or form various kinds of mixtures.[16] The early modern body was also a space in which body parts could move about freely. This is manifested both by Forman's medico-astrological rules (tying a certain state of the heavens to a specific humoral and medical condition) and by direct reports from Napier's patients. For instance, in the case of Mr. Underhill, a man from the nearby village of Willen, Napier noted that he was "somewhat lyeng [lying] on the left syde because his hart moveth in him."[17] In another case, the twenty-three-year-old patient Anne Wood complained that "she feeles somewhat move in both sides of her belly."[18] Another common disorder was the "rising" of an organ, usually the lungs, stomach, or womb, which could block the patient's airways or even harm another organ. For example, in the case of Agnes Goodman, Napier recorded "a rising in her stomacke up to her throate ready to stoppe her wynde."[19] Some of Forman's medico-astrological rules reflect this movement of organs. According to one rule, if Mercury was Lord of the sixth or the twelfth house, the patient was troubled by "many stinking cold flegmaticke humors in the matrix [womb], which do breed wynde and cause the matrix to rise to the stomacke."[20] Another astrological state signified "much payne of the head and stomacke, and a rising of the lungs into the throate redy to stop the wynde."[21]

Most important in the early modern body image were the orifices, which were seen as places of continuous exchange between the interior body and the outside world. Since the inside was envisaged as a place of invisible flow, excreted fluids were a major source for its understanding.[22] Some of the abundant or corrupted humor was discharged through the orifices, so one could identify it in the vomit or excreta and reach a correct conclusion as to what was happening inside. The nature of the fluid discharged, such as the appearance of vomit or the color of discharged phlegm, revealed significant information about the body's hidden activities and the state of interior fluids. Napier made note of such bodily discharge very infrequently but usually in great detail. For example, he recorded that a three-year-old girl from Astwood Berry "vomiteth very litle at once, . . . a kinde of shinne flegme of a waterish colour, somewhat whitter,"[23] and that a fifty-year-old gentlewoman from Haversham "spitts much flegme mixed with yellow & greene choler."[24] Occasionally, Napier also made note of the patient's urine, a very popular tool in early modern diagnosis, especially in the daily practice of irregular practitioners. I will expand on Napier's use of such physical evidence later in this chapter.

The importance of bodily excretions lay not only in their role in deciphering the hidden. Regular discharge was a prerequisite to the proper function of the body, since in a healthy body, the humors were in constant motion and superfluous, or corrupted matter had to exit the body regularly. In sickness, on the other hand, the flow of fluids was interrupted, and the channels to the outside world were blocked, fully or partially.[25] Maintaining that in sickness the body lost its natural condition of openness, Gianna Pomata argues that "the sick body was described first of all as a closed body, one that needed to be forcibly opened for the expulsion of the disease trapped inside."[26] As a result, there was a constant risk of obstruction in the body. A very common complaint among women was stopped menses, and patients of both genders often suffered from constipation.[27] The very popular practice of bloodletting, widely deployed by physicians from ancient times to the mid-nineteenth century, was also fundamentally rooted in this view of the body as in constant need of evacuation. Both natural (as in menstrual) and initiated bleedings were thought to free the body from excessive blood (*plethora*) or even to prevent the development of such an excess.[28] Nevertheless, a medical problem in Napier's consulting room could also be associated with excessive discharge. Such was the case with a twenty-eight-year-old woman from Paulus Perry (now Potterspury), who came to him because "her body opens when she goes to stool or to [make] water."[29] Other common types of excessive flow particular to women were white genital flow (*album profluvium*) and superfluous menstrual flow (*menstruorum redundantia*).[30] In fact, having too much flow seems to have been as much a problem as having too little. Health, therefore, was not defined in terms of a binary state of openness or closure but rather required a "normal level" of matter exchange between the inside and the outside.

The Image of Illness

Humoral medicine was based on a philosophical pattern deriving from Aristotle's four basic elements of the world and the conception of the body as a container of flowing and interacting humors. According to Galen, "Diseases occur when the humours decrease or increase contrary to what is usual in terms of quantity, quality, shifting of position, irregular combination or putrefaction of whatever has rotted."[31] Such "humoral imbalance" (or, more precisely, the alteration of humoral mixture) could result from external conditions, such as an unusual effort or an act of sorcery, or from natural inner processes that "went wrong." For example, food, while containing the nutriments needed to counter the natural decay of the body, also contained superfluities that had to be discharged through the body's orifices or else would become a corrupting, harmful presence. Galen explained that when an ill diet is taken, the normal functions of the body are not sufficient to evacuate the surplus and the waste matter that is trapped and putrefies and corrupts the humors.[32] The same destructive alteration of humors could be the outcome of a physical shock, such as a bruise or a sudden cold. Several recent studies in the history of medicine have shown that illness was believed to be primarily rooted in this corrupted waste and in the disruption of proper balance between the body's input and output.[33] This image of sickness and health, which ruled Western medicine from late antiquity until well into the eighteenth century, also guided our Great Linford doctor, as well as the vast majority of his contemporaries. Napier did not often use the rhetoric of humoral medicine, but once noted that in the body of one of his patients "humours disperst throughout every part seeking issue,"[34] and he did refer quite frequently to an excess of phlegm, which was one of the most common conditions of humoral imbalance.[35]

An important element in the conversation between the inside and outside was the skin: the body's natural boundary, through which it exhibits its state. Galen argued that since the skin is situated outside everything else it receives the waste from the whole body, and therefore its alterations can be used to ascertain what is happening internally. The attempts of corrupted matter to exit the body through the skin, causing local swellings, boils, and sores, are evident from both medico-astrological rules and records of daily medical practice.[36] One of Forman's rules describes a condition in which a venomous humor could "make the face swell and fingers and hands and leggs, and brakes out in pimples and blisters."[37] Other doctors attempted to explain the course of an illness by following the visual manifestations caused by bursting humors trying to make their way out.[38] Excreta, urination, sweating, and natural bleeding (i.e., menstruation) were nature's way of getting rid of trapped waste, while vomiting, diarrhea, catarrhs, nosebleed, or spitting of blood acted as "emergency" modes of evacuation. If nature completely failed to take its course, professional intervention was needed in order to achieve the same result—by purging, vomits, bloodletting, invoking massive sweating, and the like. Recovery was thus achieved, not by cautiously regaining humoral balance, but by violently cleansing the body of the trapped or corrupted matter.[39] Later on in this chapter, I will show how Napier assessed the success of his treatments through careful measurement of the bodily output (usually vomit and stools) invoked by his medicines.

The medical practice of many popular healers focused primarily on patients' immediate complaints and visible symptoms. Since these usually involved pains as well as observable bodily phenomena (such as sores or swellings) and noticeable dysfunctions, lay practitioners frequently dealt with pain relief and addressed only the visible, superficial problems.[40] Such practices were severely criticized by educated physicians, who prided themselves on their skill at getting to the root of the problem.[41] "It was the ability to interpret signs correctly," argues Duden, "that distinguished him [Dr. Storch] from the pure empiricists, who considered only superficial phenomena and did not know how to get at the hidden causes."[42] Today, the cause of disease can be practically "seen": bacteria, kidney stones, ruptured appendixes, and inflammatory tissue are disease matters that can be directly or indirectly observed thanks to various modern technologies. By contrast, early modern physicians had to work "in the dark," yet still directed their diagnostic gaze at the invisible processes inside the patient's body.[43]

According to the *doctrine of signs*, an understanding of disease and health relied on the rational interpretation of sensory data known as *symptoms*. The product of such an interpretation was a *sign*, and a set of signs implied a *cause*, which was the essence of disease. Individual symptoms had no significance by themselves in diagnosing an illness or devising a therapy; they were valuable only as sources of signs. Moreover, clusters of signs did not necessarily identify a specific disease, and diagnosis could not be "naively" performed based on symptoms and their formulaic interpretation without going through a rational process of thought. In fact, some illnesses had almost indistinguishable symptoms.[44]

There were three possible types of sources of signs: the patient himself, his attendants, and the doctor's observations. Sources of signs could be, for example, damaged bodily function, the patient's disease story, excretion, or pain. In his handbook of practical medicine, Christopher Wirtzung listed the signs of a headache caused by cold: "If the paine of the head proceede of cold flegmaticke humors and continue long, then are these the signes: wearisomnesse of all the parts, and as if al the body were beaten and broken into peeces : the paine is not extreme, without any swelling or thirst, sleepinesse, much spitting at the mouth, much moisture at the nose: for such like humidities doe daily increase in the diseased; the face is always palely coloured, & somewhat swollen, the eyes run, and the mouth is quite out of taste."[45] Since a sign could also be a symptom, a cause, or a disease, medical evidence was inevitably rather vague and ambiguous. The theoretical framework therefore allowed (or rather required) doctors to have a very flexible approach in practice. Its complexity, incoherence, and ambiguity made the *doctrine of signs* the Achilles' heel of humoral medicine. The lack of absolutely certain signs and the innate confusion of categories (cause/symptom/disease, objective/subjective, sensible/intelligible) made the diagnostic process unreliable, complicated, and even the subject of ridicule. Unfortunately, most competing methods of medical judgment, usually too simplistic or general, did not supply a better alternative. Revealing the root of the medical problem was early modern medicine's weakest link.

To make the physician's life even harder, the pain or damaged function was not necessarily situated in proximity to the diseased body organ, and so the setting of the pain or dysfunction was not always significant for making a medical decision.[46] Nor were the settings of pain and disease steady or fixed over time. As humors were thought to be flowing freely inside the body, the movement of disease was a rather reasonable inference. Andrew Wear ties the early modern narrative of disease to the portrayal of the body as an interconnected whole, arguing that "disease travelled distances across the body and often seemed to develop in stages or episodes across place and time, as in a story."[47] Although most medico-astrological rules described diseases that sat in specific parts of the body, others portrayed precise movement: "a venemouse rhewme [rheum] running up and downe the body, head, stomacke, backe and belly, sides and downe into the feet."[48] Lady Windsor, a woman of thirty-two who visited Napier in March 1610, was "very mutch swelled by fits, sometymes in her stomacke & then discending downe into her legs at what tyme her stomacke is well, but after her legs grow smale & then the humour goeth vp to the stomacke & then growth very sick & casteth somewhat & then hath a hard [cake?] on syde of her bellye."[49] Such movement of disease in corporeal space often resulted in a corresponding movement of pain. Claudia Stein quotes the sixteenth-century German physician Alexander Seitz, who maintained that, according to Galen, "a back-and-forth moving pain, is nothing other than a vapor or wind." According to Stein, Seitz understood pain as having some kind of a material essence: "This pain-vapor was rooted in the poisonous disease-matter itself. Sometimes, however, the matter would expel the vapor and would then 'chase' it throughout the body's veins. Hence the feeling that patients had of pain moving incessantly all over their body. In some cases, the pain-vapor could get stuck in very small veins and would slowly turn 'into a matter equal to that it originated from.'"[50]

Wandering pains inside the body were not simply a theoretical possibility; they were also an actual daily experience of contemporary patients. From the patients' perspective, pain was something that traveled throughout the body, changing site and form, while still remaining the same *thing*.[51] Many of Napier's patients experienced pain that had moved from one part of the body to another or that was generally "running from one place to another."[52] Agnes Hilpin, a forty-year-old woman from North Crawley, had "a payne running from place to place, sometymes in her head, sometymes in her feet, sometymes in other parte of her body."[53] In the case of Alice Goddman, Napier noted that her pain "comes up sometymes into her throte and head, very cold and chilling like an ague, going round about her body and closing at her hart."[54] In the case of a twenty-nine-year-old man from Stoke Golding, Napier recorded that "a great ache runnes about him in his chest, armes and leggs, and makes him giddy."[55] In other cases, Napier reported a more specific route of pain in the body, which was yet another element of disease history. John Hollywell, a twenty-eight-year-old man from Barsham, Suffolk, had "a great pain in his head which fell downe into his eyes and from thence being drawen backwards by applying of things to his nape went downe the reynes of his backe and into his legges."[56] William Anglissay, a thirteen-year-old boy from Marston Moretaine, "had first a payne in his shoulder and lefte syde with rednes against the region of his hart.

afterward it came into his loynes and now [he is] with great payne and shooting in his groyne and members [testicles] on the lefte syde."[57] These and other cases reveal that the movement of pain comprised a substantial component of Napier's medical narratives, suggesting that not only disease concepts but also symptoms and sensations can change over time and are culturally bound. It is a further confirmation of the view that medical cosmologies tend to shape not only the professional medical discourse but also patients' accounts.[58]

A third difficulty confronting the process of medical judgment was rooted in the complex structure of causation.[59] According to Forman, disease could originate from three possible causes: natural (internal), unnatural (also referred to as "accidental," "external," or "against nature"), and supernatural.[60] This division was derived from Galen, who maintained that "all causes of these sorts of diseases either force their way in from the outside, or else they stem from the body itself; from without come all injuries that can bruise or wound, from within come excessive and irregular movements of the body, and any bad elements in the humours that have power to corrode."[61] Nevertheless, an ill state of the humors that was caused by a malicious constellation of the heavens in birth or at a critical moment was still regarded as a "natural happening" of the body.[62] Among the unnatural causes of disease, Forman listed physical harm (hurt, bruise, straining oneself), inappropriate intakes (drunkenness, surfeit, eating out of due time, keeping an ill diet, taking of medicines), sudden temperature changes ("cold taken upon heat," "drinking being hot"), sexual misconduct (lechery, venery), intentional malevolent influence (magic, witchcraft, the action of some spirit), and intense emotions (grief, fear, love) and thoughts.[63] The third category, that of supernatural causes, he attributed only to "the anger of god on a man or on his generations fore some offence committed."[64]

The above division is very revealing of early modern categories of internal/external, natural/unnatural, and physical/mental.[65] Yet the early modern tagging of an illness as "external" referred only to its initial cause with respect to the physical body. Eventually, the affliction—be it natural, unnatural, or supernatural, physical or mental, inward or outward—was always reduced to a malignant constitution of bodily humors that caused sickness in a *physical* way.[66] A bruise or a blow, a sudden cold or inappropriate food, caused illness by disturbing the inner humoral balance. Duden explains that "[a] sudden shower could chill the blood, drive it inside, cause it to stagnate,"[67] while similar explanations were given for the influence of sexual misconduct, witchcraft, or fierce emotions. Fear-induced disease, for example, was ultimately seen as an ill state of the humors that had to be treated by evacuation.[68] According to Duden, both fright and anger affected the body physically, but in opposite ways: "Fright penetrated, drove the blood from the limbs to the heart, caused the heart to tighten, to suffocate under the abundance of blood. Anger caused the blood to surge to the periphery, toward the head, into the limbs, into the womb, where it caused cramps by its surging."[69] In an illuminating illustration of his understanding of the relationship between emotions and physical ailments, Napier writes, "this greefe taken by her husbands bad & vncourteous dealing hath stopt her courses & is the principall cause of her sicknes."[70] However, reducing the problem to a matter of humoral constitution did not necessarily simplify the task for the medical practitioner, who still had to untangle the complex matrix of possible ailments, causes, symptoms, and signs. Striving to identify the true nature of the disease from among the numerous possibilities, Napier found in astrology a reliable and efficient technique for making a medical judgment.

The Basics of Astrological Medicine

Astrology has always had a place in Western medicine, and the roots of astrological medicine lie firmly in classical texts. After a period of apparent decline during the late Middle Ages, it was revived in England (as in Europe as a whole) in a burst of publications during the late Tudor period. Many physicians, including some members of the College of Physicians of London, practiced astrological medicine or published medico-astrological treatises. Referring to the professions of astrology and medicine in the sixteenth century, Charles Webster even argues that "the two avocations were compatible and partly interchangeable."[71] The principles of astrological medicine were rooted in belief in the macrocosm-microcosm relationship and the influence of celestial objects on the organs and humors within the human body. Napier himself explained that "god as the principall wheele . . . [has set] the heavens & the starres therein as a second cause in the moving, altering & disposing of thinges beneath,"[72] and that "the heavenlie bodies make their action & impression upon these earthlie bodies, which lye here beneath them."[73] "Every kinde of disease that man hath," he further explained, "hath his starre under the which it is ingendred and increaseth."[74] Most astrological doctors understood disease as a humoral imbalance, which was caused by the influence of a malevolent planet or constellation of several planets. "Hippocratic and Galenic medicine," argues Allan Chapman, "could be practised without recourse to astrology, but an astrological practitioner tended to conceive his subject in terms of the premises of humoral physiology."[75]

The methodological rules of astrological medicine, listed in numerous printed and manuscript guides, dealt with causation, judgment, prognosis, and treatment. Diagnostic rules relied on multiple kinds of correlations between planets and humors, planets and body parts, astrological signs and body parts, and stellar constellations (usually a combination of "planet in sign," or the aspect between two or more planets) and humoral conditions. The foreseen picture of the heavens for the immediate future revealed the upcoming behavior of disease and its critical days, while specific astrological states revealed the cause of disease, its nature, and its outcome. The association of the humors with specific planets was probably added by the Arabs, but no sources survive on the justification of these correspondences. Variations in individual complexion and celestial influence at birth explained why a certain person was sick at a specific moment while others were not.

A typical guide to astrological medicine instructed the practitioner in drawing an astrological chart, in classifying the disease through every possible parameter (natural/unnatural/supernatural, acute/chronic, long/short, physical/mental, etc.), and in identifying the illness and affected body parts associated with each constellation of the heavens.[76] It also taught him the appropriate times to collect herbs and prepare compound medicines, when to administer them, and the prospects of recovery or death. The use of these guides in daily practice was based on a set of astrological tables, including the year's *ephemeris* (an astronomical almanac), a *table of houses*, and a table of the aspects of planets. These were sometimes included in the manual itself, but also sold separately on the streets. Astrological treatises, as well as printed almanacs, flooded England during the second half of the seventeenth century.[77] Most guides gave similar instructions, up to and including identification of the dominant astrological forces, but differed on methods of interpretation. Thus, the affected body organs could be deduced from the place of the significant planets, from the signs in which these planets were found, or from the combination of both ("planet in sign"). This allowed practitioners flexibility in procedural choices while still adhering to the same theoretical framework. The modern historian, attempting to reconstruct the actual professional guidelines and daily practice of such healers, is overwhelmed, however, by the vast number of possibilities. Quite paradoxically, this expansive nature of medico-astrological guides tends to obscure what astrological physicians actually did in practice, where speed, accuracy, and practical coherency served as primary concerns.

Identity and Time

Identifying the Patient

Nearly all of Napier's case notes open with the patient's full name, his or her place of residence, and current age. These three personal details are hardly ever missing in the medical records, and when they are, it is often because the patient already was well known to him (either familiar enough to be known by his or her title and last name,[78] or a child or a servant of one of Napier's regular patients).[79] A member of the gentry was sometimes identified by his title and last name only. In very few cases was the first name omitted for someone of the lower classes; in such cases, the last name was prefixed by "Goodman" or "Goodwife."[80] Still, in the vast majority of the consultations, Napier made the effort to record the patient's full name and age, for both nobility and common folk. He picked up this practice from Forman, who had taught him the basic rules of an astrological-clinical interview: "The second question is, youe shall aske the name of the partie that made the vrine and the age. [F]or by the name youe may knowe where yt be a man or a woman, by the age youe shall knowe whether he or she be younge or olde, for ther is diuers things to be considered in the discover[y] of the dizeases of men and of women and of olde folkes and of younge."[81] Forman's declared reason, then, for querying the patient's name and age was that it provided him with two of the significant variables that were needed for the application of many medico-astrological rules and that could not be drawn from the astrological chart: the patient's sex and age-cohort.[82] Napier, much less concerned than Forman with the possibility of his patients withholding authentic personal information, recorded these two pieces of data for a quite different reason: he was after a unique identification of his patients. This is confirmed by his meticulous habit of recording the patient's place of residence: a datum that had no significance at all in terms of astrological judgment and was recorded by Forman only in a minority of cases.

Establishing patients' identities was by no means the general rule in early modern medical or astrological practice, and contemporary practitioners failed sometimes to record even the most basic personal information.[83] Theoretically, astrological medicine, especially when based on the time of visit, as in Napier's case, could be practiced without any regard to the identity of the individual sick person. Since Napier seldom recorded his patients' temperaments and charted their nativities very rarely in the routine of medical practice, he had no patient-specific information to incorporate in the assessment and interpretation of medical evidence. Rather, documenting this fixed set of demographic properties (name, age, gender, place of residence) allowed him to establish a continuity of treatment in returning patients (on which I will further expand later in this chapter).

Nevertheless, the patient's gender and age, as particular items of information, were important to Napier for several other reasons. First, as I have just pointed out, many of Forman's medico-astrological rules referred to gender and age in the sense that the same constellation of heavens could entail a completely different interpretation for patients of a different sex or stage of life, as the following rules suggest:

> [Saturn in Virgo in the sixth or twelfth house; Moon separating from Mars in Taurus] Causeth payne in the head ... and in the hips, and much greef and stopping of the veines of melancholy and choler in an old body, and the greene sicknes in a young woman, and stopping of the menstruall course, and the revulsion of the course in an old body. ...
>
> [Jupiter in Gemini in the sixth or twelfth house; Moon separating from Mars, which is 28 degrees in Taurus] Causeth in an elderly woman paine of the backe, head, side and belly, by revulsion or stopping of her course, and a scowring or laske in a younger body about 30 and a fever.[84]

The significance of the patient's age lay not only in the obvious biological evolution of the human body and its relation to health and disease, but also in the more general framework that correlated each of the four humors with one of the four stages of human life: childhood, youth, adulthood, and old age. Gender was particularly important in the context of women's problems, to which the physician had to give special attention.[85] Complaints such as menstrual disorders or a genital flow gave Napier further information when assessing the clinical situation, and at least in some of the cases the patient had to be asked directly for this information. Another pertinent issue was the question of pregnancy. Although the vast majority of Napier's patients posed medical queries, some young women came to Great Linford specifically to inquire whether they were gravid or not.[86] In fact, the question of pregnancy was hanging over any consultation with a woman of the relevant age, especially (but not limited to) married ones. Since pregnancy was a very common physiological condition of the body and implied certain known symptoms, it was necessary to eliminate or consider it in order to avoid unnecessary diversions in the search for the underlying medical problem.

The Significance of Time

Napier used a branch of astrology known as *horary astrology*, in which calculations are based on the exact time of the question. This method is rooted in the belief that each question (medical or other) has its own "time of birth," on which an astrological chart can be erected and conclusions drawn, just as in *natal astrology* the nativity of a person is based on his time of birth. Such a method had considerable advantages as compared to basing the chart on the *decumbiture*: the time when the patient first "fell ill" or "took his bed."[87] First, it was not always clear when that moment had occurred, especially in chronic diseases. Second, in a time when only a few people kept a watch, any estimate was bound to be inaccurate.[88] The exact time of the question was therefore the foundation of all astrological calculations that were to follow. "The first thing to be noted" in a medical consultation, declared Forman, "is the instante tyme when the question is made or the vrine brought."[89] To avoid disastrous mistakes and in order to make the entire course of the consultation replicable (in the sense that the same conclusions could be drawn again for the same case), it was essential that this time be correctly and carefully recorded. Of course, determining the "correct" time of the question was not as easy as it might sound, since real-world scenarios were always more complicated than any set of theoretical rules. Very often, a significant time lapse existed between the moment in which the urine (or query) was sent by the sick person and the time that it arrived at Great Linford. For example, in the case of Mr. Underhill, who sent one of his servants "for a purge," Napier noted that "he sent yesterday," wrote the exact time at which the servant set off, and erected the astrological chart for that time rather than for the time of his arrival.[90] However, in most cases he resorted rather conveniently to the time the patient arrived at his door. If several patients came at the same time but were treated separately, the time of arrival was still the time by which the chart was drawn.[91]

Napier also frequently recorded the duration of the illness prior to the consultation,[92] and on rarer occasions, also the exact moment at which the patient fell sick, although he still erected an astrological chart for the time of visit rather than for the decumbiture.[93] Less often, he made note of the ailment in reference to other astronomical phenomena, especially if its behavior was determined by the phase of the Moon, which was considered to have a significant influence on the human body.[94] Since all corporeal conditions were supposedly paralleled by the stars, both disease and cure corresponded to the specific state of the heavens at each specific moment. I will return to this significance of timing in later sections.

Collecting the Evidence

Reading the Heavens

In the vast majority of Napier's case notes, identity and timing coordinates are immediately followed by an astrological chart (figure), which is visually the most prominent component in his documented consultations (see fig. 2). In some cases, a single chart was drawn for several adjacent cases, while on rare occasions more than one figure was erected for the same visit. The consideration of celestial influences began even earlier in the consultation, when Napier noted the planets that were *Lord of the day* and *Lord of the hour* next to the consultation date and time.[95] These two significators could be located in a table designed "to know what planet rules every houre either of day or night," which appeared in many almanacs.[96] Napier almost always noted the first, and only occasionally the second. The figure itself was drawn of a square divided into twelve triangles by two sets of parallel lines. The dozen triangles represented the houses of the zodiac: the middle-left representing the first house (the ascendant) and from there going counterclockwise through to the twelfth.

The chart was erected in three main steps: calculating the *sidereal time* for the moment of the question, placing the zodiacal signs on the cusps (borders) of the houses, and placing the known planets, the Moon, the Sun, and the Dragon's head and tail,[97] inside the houses.[98] The task could only be accomplished using two kinds of astrological tables: the year's *ephemeris* and a *table of houses*, both of which had to be on Napier's table during encounters. The *ephemeris* gave the daily positions of the planets for each day of the relevant year, while the *table of houses* specified the borders of the houses for each *sidereal moment* within the twenty-four-hour day. The latter, built according to the Regiomontanus system,[99] was calculated for a specific latitude and was not year-specific.[100] Napier probably used a table that was designed for the latitude of London.[101]

In order to find the *sidereal time* for the time of the question, Napier calculated the time from the previous noon to the exact time of the encounter. He then opened the *ephemeris* at the page corresponding to the current month and checked the place of the Sun for the day of the consultation (or for the previous day, if the visit was held *ante meridiem*). This showed him how far, in degrees, the Sun was into a specific zodiacal sign at noon that day. Next, he opened the *table of houses* at the page dedicated to the Sun in that sign, went down the second column (titled "the 10th house") to find the aforementioned degrees, and noted the time that appeared in the first column ("time from noon") for this row. This constituted the *sidereal time at noon* for that particular date. Napier noted this value next to the still empty figure, wrote under it the exact time of the encounter, and then added the two.[102] If the result exceeded twenty-four hours, he subtracted twenty-four hours from it and corrected the date in his calculations to the following day. The result was the *sidereal time for the moment of the encounter*. Next, he turned back to the *table of houses* and looked for the row in which the value in the first column was the closest to his result. The values in that row gave the exact place in heaven for the six left houses (the tenth, eleventh, twelfth, first, second, and third), specifying how far, in degrees

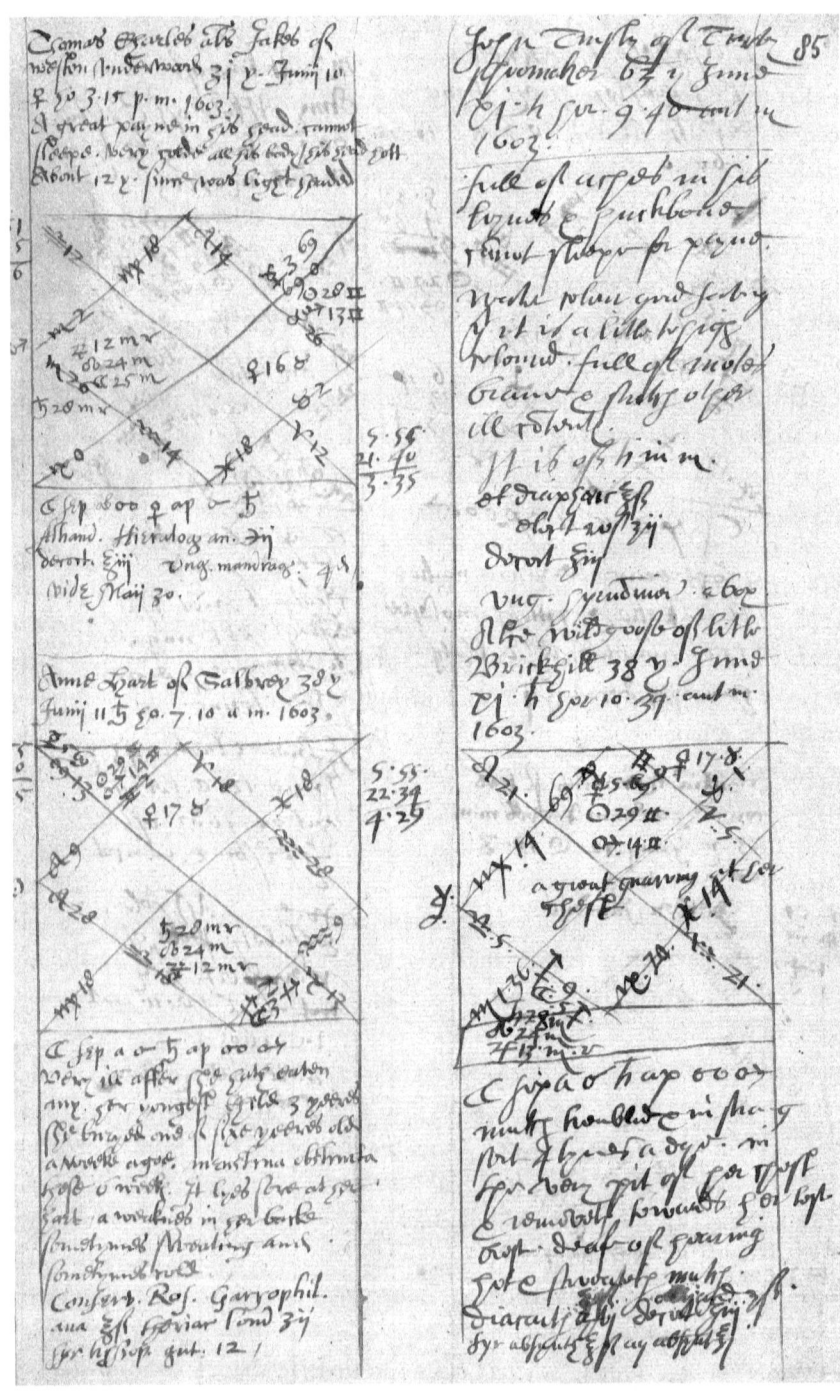

Fig. 2 Entries in Richard Napier's casebook for June 10 and 11, 1603, neatly recorded in Gerence James's hand. Ashm. 207, fol. 85. By permission of the Bodleian Library, University of Oxford.

and minutes, the house started in a specific sign of the zodiac. Napier indicated these values near the cusp of each house (usually the sign and degrees only) and then did the same for the remaining six houses, which exactly mirrored the former (same degrees in the opposite sign). Although in principle each of the houses consists of exactly thirty degrees, in practice, in this kind of astrology, the size of the houses varied slightly and was recorded by this operation.

Lastly, Napier had to place the known planets (Mercury, Venus, Mars, Jupiter, and Saturn), the Sun and the Moon, and the Dragon's head and tail inside the houses. In order to do so, he went back to the same row he had consulted before in the *ephemeris* and found the places of the planets in the zodiac for that day (how far in degrees and minutes, and in what sign). He then placed each planet according to the borders of the houses; for example, if Leo 8° was on the cusp of the sixth house, Virgo 3° on the cusp of the seventh, and Saturn was in Leo 23°, then Saturn was placed inside the sixth house. If Saturn was in Leo 5°, then it was placed inside the fifth house, because its place in heaven lay between the two points marked by the borders of this house. If again, Saturn was in Leo 23°, but the cusp of the seventh house was Leo 22° instead of Virgo 3°, then the planet was placed inside the seventh house. After the chart was drawn, Napier noted to its left the *Lord of the Ascendant* (i.e., the planet that rules the sign found on the cusp of the first house) and underneath it the *Lord of the hour* (i.e., the planet that signified the current *planetary hour*).[103] Mapping the sky up to this point was rather a straightforward procedure, sometimes complemented by a more extensive astrological analysis. For the patient, all this appeared as an obscure procedure, involving odd graphical symbols that signified meaning in a way he or she could not begin to understand. For Napier, the completed chart and following calculations were a symbolic representation, not only of the inner body state, but of the whole medical situation, including its past, present, and future. Through the symbolic, technical language of astrology, the chart acquainted the doctor with the facts of the case, from which meaning was yet to be drawn.

Considering Other Signs

In Napier's consulting room the astrological chart was undoubtedly the primary source of medical evidence, but it was by no means the only one. If it had been, it would have made his procedure completely impersonal. Typical medical assessment in the Renaissance included noting the patient's manifest symptoms, listening to his or her narrative of disease, and inspecting excretions.[104] Napier, being no exception, utilized these three sources of information to map what he had concluded from the astrological chart against various corporeal coordinates.

Although full-scale physical examination began only in the nineteenth century, at least some kind of direct medical observation was widely deployed in medical practice much earlier, and early modern doctors often noted visible injuries and sores, swellings and boils, changes in skin color, handicaps and malfunctions.[105] Napier himself conducted some kind of physical examination, at least in some of the consultations. When Dorothy Coxes, troubled in her "backe, wast [waist], shoulders, right arm & leg," returned to him after a week, he noted, "her backe, belly and about her wast [waist] smarteth and nothing to be seene."[106] Obviously, some kind of visual observation was needed to reach that conclusion. Other types of inspections are implicitly referred to in the case notes. For instance, in the case of the fifty-year-old William Southwicke, Napier remarked that the patient was "very sore in his body when he is touched or stirred."[107] Such a remark naturally implies that some kind of physical contact between patient and doctor took place. In many other cases, Napier noted visible swellings and sores,[108] changes of skin color,[109] visible attributes of the patient's blood,[110] and even inflamed tonsils.[111]

One of the most popular types of physical examination in the seventeenth century was that of taking the patient's pulse. In Napier's case, only rare records exist of such a procedure.[112] On the other hand, body temperature was considered a significant sign and was frequently recorded.[113] Heat and cold, two of the four basic attributes of nature and of humoral medicine, were important, not as mere symptoms, but as supporting signs of the corrupted or abounding humor. Napier noted not only exceptional body heat, but also differences in temperature between body parts ("very colde all his body & his head hott"),[114] changes in temperature over time ("very cold & chill & presently, after hott & burning"),[115] and how body temperature affected the illness ("he is well when he is hott with worke but very ill when he is cold").[116] Still, the physical assessment of patients played only a minor role as a source of information in Napier's medical judgment, and there is no evidence to suggest that it was conducted regularly in his consultations.

Another bodily examination conducted by many early modern practitioners was urine inspection (uroscopy): the rough inspection of the patient's urine.[117] Based on the assumption that the corrupted humors could be revealed through the body's excretions, uroscopy had become so widespread by the sixteenth century that it was rarely absent in any medical practice. However, as its popularity grew, urine inspection began to be severely challenged by learned physicians and medical authorities and gradually became the definitive mark of the medical charlatan when used as the sole means of diagnosis.[118] Forman was very skeptical about urine inspection and voiced his opinion at length, declaring that since the color of urine could change as a result of food, body temperature, or the effect of some humor, it could not be used as a reliable tool in medical judgment.[119] Lauren Kassell maintains Forman saw the flux of urine not as an indicator of disease, but as merely "a token of the patient's will to be healed."[120] Napier, taking as always a less extreme approach than his tutor, occasionally took the trouble to consider his patients' water. Nevertheless, his short remarks, very often given in Latin, mostly describe the urine's physical features, failing to offer any consequent interpretation. These remarks typically provide a very brief account of the urine's color (e.g., "red," "yellowish," "pale," "high coloured," "black"), texture (e.g., "muddy," "thick as a puddle"), contents (e.g., "full of rags," "a great peace of matter in her water"), or looks (e.g., "good urine," "shining," "clear as well water," "full of bubbles," "with yellow froth"). Most of these features were referenced in medico-astrological rules, as demonstrated below, allowing Napier to double-check his medical judgment. Only infrequently did he conduct a more thorough examination, as when he reported a "thicke white water, when it had stood a white creame thine [thin] over it, when it was heat[ed] the creame grewe thicke and tough."[121] Sawyer has already pointed out that early modern patients expected their doctors to look at their urine and that their expectations comprised one of the reasons for the use of urine inspection in Napier's consulting room.[122] Yet although in his earliest regular practice book (from 1598–99) urine descriptions are quite common,[123] case notes dating only a few years later reveal a significant downturn in the frequency of their use. This decrease in the inspection of urine could be due to a heightened sense of professional confidence or, as he admitted years later in front of two clergymen, because the assessment of urine had deceived him far more than astrology.[124]

The Voice of the Patient

Medical case notes naturally reflect the doctor's standpoint and sum up what is captured by his professional gaze. This is especially true for premodern medicine, in which the distinction between subjective and objective perspective did not yet exist.[125] Restoring the patient's voice, muffled by the doctor's interpretation and selection processes, is thus one of the trickiest tasks for the modern historian. The issue should nonetheless be tackled, as patients' narratives inside and outside the consulting room often played a vital role in influencing the doctor's medical judgment. In her study on early modern "French pox," Stein argues that the opinion of the sixteenth-century pox sufferer and his disease narrative were both important to the hospital's educated examiners and that diagnosis relied equally on both.[126] Was this also the case in Napier's consulting room? His mentor Forman seems to have objected to such practice: "for a man by astrologie wille saye more by a question demanded for the state of the sicke, for his sicknes and diseas and totchinge life and death or cueringe of the party hurte or sicke," he claimed, "then ten phisisions that shall see the vrine or speake with the sicke bodye." Yet the continuation of this passage reveals that the astrologer was referring only to patients' self-interpretation of their disease, "[f]or many ar sicke that knowe not their owne dizeas nor the cause."[127] Napier himself constantly documented his patients' complaints: initially in the form of a short list of pains or sensations, and later as descriptive short narratives. Some descriptions were very vivid,[128] while others recorded only a vague list of organs. Sawyer chose to emphasize this "dialectical interaction" between Napier and his patients, arguing that in many cases his judgment was based on the clinical interview rather than on the astrological chart and that, whenever possible, the former was prioritized over the astrological data.[129] As I will later argue, this claim might need some rethinking in light of a reconsideration of the actual source of Napier's lists of pains and complaints.

So what role did patients' complaints play in Napier's overall judgment? Before this question can be addressed, some preliminary points need to be discussed. First, as I have already noted, the organs in which the patient experienced pain or dysfunction were not necessarily the setting of disease itself. This is true for early modern medicine, in which corrupted humors could travel almost freely throughout corporeal space, carrying pain from one place to another; for astrological medicine, in which the "triplicity of signs" linked triplets of zodiacal signs to allow a fundamental connection between distant body parts;[130] and for modern medicine, where a disc rupture in the lower back can cause severe pain or neurological dysfunction in the feet. Second, in Napier's medical notes, the doctor's observations and the patient's accounts were inextricably intertwined. The task of distinguishing between the two is especially difficult because, in the early modern consulting room, patients' attested feelings were too often incorporated into the medical framework along with observational givens.[131] For example, John Hall, Shakespeare's son-in-law, who practiced physic in Stratford-upon-Avon around the same time as Napier, reported that one of his patients, "being troubled with trembling of the arms and thighs, often felt vapours ascend to the heart, thence to the throat, and often thought herself suffocated."[132] Clearly, the experienced feelings of the patient were translated here into the language of humoral medicine in a way that makes historical analysis a tremendously difficult task.

Social histories of early modern medicine, particularly in the 1980s and 1990s, emphasized the role of patients in reading their own afflictions and in negotiating what they understood to be the appropriate treatment.[133] Based on the unstructured medical marketplace of seventeenth-century England and his examination of Napier's records, Sawyer has offered the image of "the patient shuttle" for the active role taken by patients in their quest for good health. More recently, Alisha Rankin has described the attempts of Elisabeth of Rochlitz, Duchess of Saxony, who repeatedly replaced her sources of medical help in the search for what *she* saw as the appropriate diagnosis and treatment of her ailment.[134] Napier's patients might have taken an active role by shuttling through several points of care,[135] but only seldom did they arrive with a specific request. One such example is that of Alice Cutbert from Olney, who returned to Napier after receiving a treatment for a cataract of the eyes, and stated she "desyres a vomit."[136] Napier gave her antimony, which was considered a very strong emetic. In another case, the parson of the nearby parish of Simpson "sent for a potion to purge him."[137] More often, Napier simply echoed the concerns and anxieties expressed by his patients or their attendants, the vast majority of which refer to a suspected problem such as "jaundice,"[138] "impostume" (abscess),[139] "chest worms,"[140] or witchcraft.[141] Concerns of the latter sort were occasionally accompanied by a specific accusation against the person or persons allegedly responsible for the act of sorcery. Nevertheless, Napier rarely commented on his patients' suspicions one way or the other, and seems not to have taken them into account at all while assessing their medical case. This is not surprising, since, as I have already mentioned, Forman was very clear that such suggested diagnoses had no value.[142] It is notable, though, that Napier made the effort of recording these concerns, and did not explicitly counter or dismiss them.

An additional issue pertinent to the discussion of the relation of patients' complaints to Napier's judgment is that the source of what seems to be direct patients' input—Napier's list of complaints and pains—is not in fact self-evident. Almost every handbook of astrological medicine included a table that indicated which parts of the body were ruled by each combination of "planet in sign." These tables were not identical, but usually resembled each other, since they were not arbitrarily constructed: the planet in its own sign was believed to affect the head, in the next sign the throat, and so on down to the feet.[143] Theoretically, once the ruling astrological forces (i.e., ruling planets, aspects, etc.) were established, the physician could infer the whereabouts of pain and disease from the astrological chart with the help of this table. Furthermore, many of Forman's astrological rules included a list of affected organs and even sensations and pains that were attributed to a specific constellation of the heavens.[144]

Napier himself referred in one of his tracts to a medieval treatise on astrology and medicine that declared how, by using astrology, the physician "may knowe the place where the disease lyeth."[145] Were the lists of "complaints" in Napier's case records voiced therefore by his patients, or were they methodically drawn from the map of the heavens? Kassell has already suggested (although very cautiously) that in Forman's case, at least, the latter is a real option.[146]

Certainly, one thing that immediately strikes the eye on looking at Napier's consultations is the unusually long lists of affected body organs in each of them.[147] A significantly large portion of Napier's patients experienced pain at three or more separate body parts, and some ached in as many as seven or eight places. It seems rather extraordinary, at least from our modern point of view, that so many patients suffered from so many afflictions at once. Moreover, some of Napier's notes include a list of organs without an account of what was actually wrong with them. For example, in the case of a thirty-one-year-old patient from Shutlanger, Napier listed "his hart and stomacke and sometymes his ribbes," without stating the nature of the complaint.[148] This fits the language of the tables mentioned above specifying a list of affected organs that are not necessarily the seat of pain or that manifest external symptoms yet nevertheless host the diseased matter itself. Sometimes Napier's choice of words seems to imply this even more clearly—for example, in observations such as "tooke a great cold, about 6 weeks ago, lyes in his chest and backe," "riseth up in her stomacke and head," "they ly about her hart & stomacke," or "it resteth in his rayns [kidneys] & chest & sydes."[149] In other cases, such as that of Ellen Brinklowe of Newnton, Napier listed two separate sets of involved body parts: "A payne in her head and at her heart, a very weake stomacke, lyes in her joynts."[150] The choice of words suggests a distinction between the physical sensations communicated by the patient (in her head, heart, and stomach) and the setting of the ailment drawn from the chart (in her joints). Another indicative example is the aforementioned case of Mr. Underhill, who sent his servant to Great Linford to get a purge. Napier indicated "lightnes of head, abundance of flegme, feeblenes of stomacke ventosity, and almost a generall ache through all the body, but esp[ecially] in the head worke and belly. hath somewhat lyeng on the left syde because [of] his hart which moveth in him much, and grievous[ly] sighing with his great payne."[151] Could this detailed description have come entirely from the servant? I think this at least doubtful.

Yet in none of the cases I have surveyed does the array of reported body parts in Napier's medical record perfectly match the list of affected organs found in Forman's tables for the appropriate astrological state. For instance, in the abovementioned case of the patient from Shutlanger, the disease was caused by "Mars in Aquarius," which, according to Forman's tables and rules, indicated a problem in the legs. The other important signifier of disease, "Moon separating from Saturn in Sagittarius," indicated a problem in the feet.[152] In fact, none of the planets in this patient's chart were in a position that could indicate a problem in the heart, stomach, or ribs, the body parts mentioned by Napier.[153] Generally speaking, the correspondence between the lists of complaints in Napier's notes and the affected body parts in Forman's (or others') astrological guides is so poor that it cannot support any claim of compatibility. Moreover, a consideration of the symptoms recorded in cases of returning patients indicates that at least some of the problems were voiced by the patient himself. For example, in the case of the twenty-year-old Bettridge Robinson, Napier recorded that she had "a great payne on her chest [and] about her heart, a rising in her stomacke ready to stop her throate." When she came back eight days later, he recorded, "a payne at her heart, feares an impostume, very short-wynded. when she is well her belly troubled and she casts much water."[154] The two case histories are too similar to be taken from the two separate and very different charts that were erected for the two visits, but too unlike to be copied from one encounter to the other.[155] Obviously, the patient's words were taken into account here. In some cases, a detailed description of the pain is given that seems to be too personal to be deduced from the chart, such as "first a pulling, after that followes a stitche, and lastly fitts of swounding [swooning]."[156] On other occasions, Napier's choice of words is even more indicative of subjective feelings, as in the case of a thirty-five-year-old man from Emberton, for whom he noted, "when he coughes he feeles a pricking in his side."[157]

Overall, the fact that narratives of pain settings are given by Forman's astrological rules does not necessarily imply that they were extrapolated from these rules in Napier's everyday practice. In fact, even in modern medicine, disease narratives frequently discuss pains, yet serve only as a pattern against which patients' attested feelings are compared. The evidence is therefore inconclusive. My understanding is that the lists of complaints in Napier's records should be carefully analyzed, as some were voiced by the patient while others were deduced from the multitude of astrological rules he consulted.[158]

Making Judgment

Identifying the Problem

Once the map of the heavens was drawn, other signs assessed, and the patient's complaints voiced and noted, it was time for Napier to interpret the data and identify the medical problem. This was by far the most complicated and crucial step in the consultation. Because medical problems were discussed in terms of humoral imbalance or corruption, the objective of the diagnostic process was to identify the state of humors from which therapy would follow. Napier's medical judgment relied mainly on interpretation of the astrological chart, which was checked against any additional information that he had managed to gather. The first step in interpreting the chart was identifying the *astrological cause*, which was defined by the position of the planet that was *Lord of the sixth house*[159] (or of the twelfth house, if the Moon was waxing) and by the position of the planet from which the Moon last separated:

> In all your Judgments for the Party diseased, you must have an especial care and eye to what Planet the ☽ is conjoyned in full Aspect, and from what Planet she did last separate, and to what Planet she doth next apply; because that Planet to whom she doth apply, doth also shew, as well as the Lord of the 12th house in the New ☽, or the Lord of the 6th in the wane or decrease of the ☽, what Humour doth reign or is predominate in the Body, and what Humour doth increase or decrease.[160]

If, for instance, the Moon was on its wane and the *Lord of the sixth house* was Saturn, which was found in the house of Cancer, and the Moon had last separated from Mars, which was found in the house of Sagittarius, then the *astrological cause* would be "Saturn in Cancer; Moon separating from Mars in Sagittarius."[161] During his first years in practice and especially up to 1600, Napier noted the *astrological cause* under almost every chart he drew. Like a skilled mathematician, leaving out trivial steps in a long proof, he gradually neglected that habit and indicated the cause only infrequently.[162] The *astrological cause* revealed the exact constitution of the humors in the body, in all their diversity and colorfulness, but was particularly important in identifying the abounding humors: "And the 6 h[ouse] and his L[ord] in the last half of every ☽, ... shall signify the cause of the disease and hummor that aboundeth, whether it be of melancholy, choler, flegme or of the bloud, or of more hummors then one, ... and which hath most dominion, and in what degree the hummors predominant are, either in the first deg[ree], 2, 3, or 4 deg[ree]."[163]

The specific condition of the humors with which each *astrological cause* was associated was specified in many astrological guides that circulated in manuscript and in print during the seventeenth century. A reference to a book holding such rules in Forman's astrological notes suggests that he himself collected or designed such a set of instructions.[164] Still, those sporadic rules that have survived in his casebooks provide us with an idea of how he drew his conclusions from the *astrological cause*:

> Suppose it is in the last part of the ☽ and ♃ in ♈ is L[ord] of the 6 h[ouse], heere the humor that doth abound is yellow choller mixed with the bloud, a sharpe yellow chollericke humor. Then we looke to the ☽, and the ☽ doth sep[arate] from ♂ in ♋, which causeth a kinde of salt flegme and water, and the ☽ doth ap[ply] to ♄ in ♍ which causeth a strong melancholicke humor. Now if yow will knowe whether the humor that is caused by ♂ in ♋ or of ♄ in ♍ is mixed with that humor which ♃ causeth in ♈, then shall yow looke how farre the ☽ is sep[arated] from ♂ and how farre she hath to go to ♄. And if the ☽ be not gone above 4 deg[rees] from ♂ then the humor that ♂ causeth is mixed with that choler which is causeth of ♃ in ♈.[165]

In the waning phase of the Moon, the cause was determined by the planet that was *Lord of the sixth house* and the sign it was in, in this case Jupiter in Aries. Jupiter signifies blood, and Aries signifies choler; hence the combination of both signifies thin yellow choler mixed with blood. Mars is choleric and in particular signifies red choler. Cancer is a watery sign that signifies salty phlegm. Hence, the combination of Mars in Cancer signifies red choler and a kind of watery and salty phlegm. Saturn and Virgo are both melancholic, hence their combination signifies an abundance of melancholy.[166] The exact place of the Moon, and whether it was closer to the planet it had last separated from or to the planet it applied to, determined which of the two signified conditions was dominant in the patient's body, together with the condition signified by the *Lord of the sixth house*. For example, if the Moon was not above four degrees from Mars, then the disease would be "thin yellow choler mixed with blood; and red choler with watery and salty phlegm." The *astrological cause*, visually displayed as a short array of planets and zodiacal signs, thus served as a symbolic representation of the state of the inner body.

Because the nature of the medical problem lay in the abundance of malevolent humors, identifying their constitution in the body was the essential and ultimate product of Napier's medical judgment. Still, the technique described so far was only the "main route" to the identification of disease and by no means constituted the sole consideration of evidence. For instance, Napier often considered other major astrological forces, especially ruling planets, and judged the situation by their positions and mutual aspects.[167] His copy of Forman's astrological book contains an additional set of rules that listed complex combinations of astrological causes, their affected body parts, related complaints, and the state of humors they revealed.[168] Other rules referred to the position of the Dragon's head or tail and to the *Lord of the day* and *Lord of the hour* that Napier noted near the date and time of the encounter.[169]

Interpreting the chart was not always straightforward, nor was it a "closed matter." Always attempting to improve his professional skills, Napier sometimes returned to a previous visit and added new significators and aspects that would allow him to correct what he now thought to be an earlier erroneous judgment.[170] Moreover, the procedure just described was relevant only for cases in which the disease was believed to be natural (i.e., of an inward cause). A preliminary interpretation of the chart revealed the category of disease, so another route could be taken for other situations. Forman himself gave specific astrological rules that could help the physician identify the type of illness that troubled the patient.[171] The significance of a different disease category was not in choosing a different route for cure, as we might think. Rather, it just meant choosing other significators in the chart to reveal the underlying condition of humors:

> The signifier of the sicknes after the opinion of some is the 6 h[ouse] and the pl[anet] in the 6 h[ouse], which signify the sicknes comming of any inward or naturall cause. But the signifiers of the sicknes that comes by accident is the L[ord] or dispositor of the 6 h[ouse] and the malevolent pl[anet] which doth afflict the ☽ or the ☉ or that doth afflict the L[ord] of the asc[endant], or from whome the L[ord] asc[endant] did last separate or is in a full ☌ or aspect withall at that tyme of the question.[172]

Medical astrology offered many other rules of judgment that could be used to reveal the diverse aspects of the patient's condition, such as whether the disease would be short or long and whether life or death would prevail. Napier, adhering to his practical priorities and with a long line of anxious patients at his door, rarely took the time to consider these issues at all.[173]

In order to verify his astrological conclusions and avoid the consequences of miscalculation or erroneous interpretation, Napier often considered the results of physical observation, urine inspection, and voiced complaints. The color and thickness of the urine could reveal the mix of abundant humors, although not as accurately as his astrology. For example, if the urine was grayish, white, or blue, then the patient was probably suffering from a superfluity of melancholy, while reddish urine could result from the blood being mixed with melancholy.[174] Other rules correlated a certain humoral condition to data that could be gained through physical examination or questioning of the patient. If, for example, "Melancholy rest in the Sides," declared Richard Saunders, a writer of astrological medicine from the later seventeenth century, "they will have rawness and much windiness, sharp belchings, burnings, and grief of the Sides; also the Sides are plucked upwards, and many times they are troubled with inflammations, especially when Choler is mixed with Melancholy."[175] Napier could have used this rule conversely, to conclude that a patient with pains and inflammations in his sides might be suffering from an abundance of melancholy and choler. A rare disclosure of the role of patients' sensations in Napier's medical judgment can be found in a letter he wrote in January 1623 to one of his noble patients troubled with a voiding of blood:

> our booke[s] doe acquaint vs that the voyding of blood may come many ways: eyther by the breaking of a vayne, or opening of the vaynes by some cholerick humour, or by blood Issuing from the liver or splene or kydneys. now if it had proceeded from the splene, then you would by all likely have felt now & then some payne vnder the left syde, but if from the liver then on the right syde vnder the liver, if from the lungs then you would have bene provoked to cast or to spit blood & have felt a heavynes & payne about your chest & stomacke. but of any of these I doe not heare that you doe make any complaynt, but most specially of your backe, & therfore [I] doe thenc collect that it is most likely that it should proceed from your kydneys, eyther by some grating of a stone, or by some opretion of the vaynes.[176]

For Napier, this patient's disease story helped build a sound medical narrative, reinforcing one possible interpretation over another. The personalization of illness in Napier's consulting room was in fact chiefly materialized through such referral to the patient's perspective, rather than to his or her complexion.[177] Certainly, Napier did not *base* his conclusions on such superficial data as sensations or symptoms, but they did help him stay on the right track and avoid unnecessary mistakes.

What Was Disease?

Different periods and cultures have different sets of "available diseases" and, more importantly, differ in what constitutes the identity of these diseases. Traditional historiography of medicine saw disease as a permanent entity, and modern disease concepts were "extended backwards in time" to capture the medical concerns of past generations. This view was termed "retrospective diagnosis" because it assumed that diseases can be retrospectively diagnosed on the basis of texts and medical records, even if these texts and records expressed a conception of disease very different from our own. A more recent way of understanding the history of disease is through the study of theoretical disease concepts (i.e., diseases as thought entities) and their transformations over time. This view portrays contemporary disease entities as culturally determined and as mental constructs that constitute only one option among many possible explanations.[178] Other historians, such as Andrew Cunningham, claim, however, that disease cannot be merely understood as a "theoretical idea" and should rather be investigated through the medical interview or through other procedures that were used to identify an illness. They suggest that we should turn away from both retrospective diagnosis and the poststructural idea of studying the history of disease concepts and focus instead on the actual process of making a diagnosis, that is, on the people thinking and acting in a specific culture, time, or situation. The only identity that disease has, they argue, is an *operational identity*.[179]

Today, medical doctors around the world use the International Classification of Diseases coding system to define, encode, and communicate medical afflictions, and some countries enforce regulations for its use in all clinical encounters.[180] To a large extent, we regard such classification as the core of modern medical nosology. In the early modern period, disease was discussed in terms of its *causes* and *signs* and was referred to either by name or, if it did not have a recognized name, by the organs affected and its cause or primary signs.[181] But what exactly did the term "cause" refer to? Can it help us understand the early modern concept of disease? In what follows, I will briefly discuss the historiography of disease concepts and, using the notion of an *operational identity* of disease, offer an alternative nosology that is inspired by Napier's practice.

The term "disease" usually refers to a pattern of signs that is recurrently recognized in separate cases and individuals. This pattern must be deemed painful or disabling in the relevant culture and, at the same time, deviate from either the statistical norm or some idealized condition.[182] Whether these patterns are "real" or not is still a matter of dispute in philosophical and historical discourse. Half a century ago, Lester King argued in the spirit of positivism that clinical symptoms are organized in reality in predefined patterns that are no less real than the symptoms themselves. Poststructuralist outlooks, on the other hand, regard these patterns as mere conceptual constructions. Still, both approaches accept the notion that the classification of sickness is based on recurring arrangements of symptoms. Mapping the landscape of Napier's afflictions in these coordinates, one will find the following among his most common diseases: yellow jaundice, green sickness, consumption, bloody flux, pleurisy, various kinds of fevers, and wind colic.[183]

Most historians of medicine today understand early modern disease to be a definable medical condition that manifested itself through a collection of distinct symptoms, which were its components. Each of these components could also be a medical condition in its own right and could participate in different collections of symptoms that constituted other diseases as well.[184] However, conceptualizing early modern medicine in this particular way leads to some unresolved difficulties. Investigating the French pox in sixteenth-century Augsburg, Stein found a high level of inconsistency between this disease and any fixed set of symptoms and suggested that the innumerable different complexions that were exhibited by individual patients resulted in hundreds of different species of the pox. Since the pox in one's body could alter over the course of time, even to the extent of becoming a completely different disease, she claimed that "no early modern person perceiving the pox understood it as etiologically, morphologically and symptomatically separate from other disease entities." Stein concludes that disease definition in the early modern period was fluid, flexible, and temporary, with only the underlying metaphysical concept providing some constancy.[185] I find this a somewhat disturbing result because it seems to undermine the ability of early modern medicine to provide a coherent and firm theoretical framework for practical purposes. It should, I believe, encourage us to question our underlying assumptions in exploring early modern disease.

The term "cause" carried three distinct meanings in early modern medicine. The first referred to the event that was responsible for the emergence of the medical problem: the cause after which the disease has followed as an effect. An unnatural (or external) illness always had an external trigger, a *reason* for its occurrence, and references to such causes are widespread in both medical and astrological treatises.[186] Early modern disease was typically multicausal. This meant that each medical condition could result from several possible chains of events, none of which undermined the reality of the other. Thus, contemporaries worked with a complicated multifactorial structure of causation that was not primarily ontological.[187] Napier often specified such triggers for his patients' afflictions, among which were counted sudden changes of temperature or unusual cold,[188] fierce emotions or witchcraft,[189] and all types of physical harms, including a fall, being hit by a stone, or straining oneself at work.[190] Any given disease could theoretically result from several possible triggers, but this does not mean that a specific condition had more than one cause.[191]

In astrological medicine, however, the term "cause" was also used to denote the significant astrological forces that were the telltale signs of disease and their place in heaven.[192] Astrological medicine was based on the notion that mental and physical ailments were mirrored in the motions of the heavens, and the astrological chart was designed to help the physician exploit the stars in order to unravel the true nature of the affliction, not its causal sources. Forman referred therefore to the forces of heaven as "natural causes of disease" only at birth, rather than at any point of one's life.[193] The phrase "It is caused of *planet in sign*" was only a symbolic means of indicating the medical problem through its reflection in the heavens, not a verdict of causation. The chart did not say *why* something happened to someone, or why it happened at all; it just revealed the "matters of facts."

The third and most intriguing meaning of *cause*, used in both astrological and nonastrological early modern medicine, referred to the constitution of humors inside the body. My claim is that the cause, in this particular meaning, was the essence of the medical problem as well as the basis for its practical classification in the context of astrological medicine. The significant difference between premodern and modern medical views here is that the former saw sickness and health as falling on a continuous line rather than as two binary opposite conditions. Some early modern medical conditions (such as certain kinds of humoral fluxes) were even "cases with no disease," in which the symptoms simply revealed medical situations that were not ill states per se.[194] Also notable is the fact that astrological medicine, especially the particular kind practiced by Napier, revealed a potentially malevolent condition in every body at any given moment.

The essential task of the astrologer-physician was not to determine the disease that best fitted his patients' manifested symptoms, but rather to reveal the underlying condition that caused their suffering. It was the humoral imbalance or corruption that Napier was essentially after and, as I will show later in this chapter, was the key to his course of cure.[195] Brian Nance, relying on the classical work of Knud Faber, argues that although the humoral imbalance displayed some regularities, the uniqueness of each patient and of each humoral condition prevented the creation of solid disease categories.[196] However, since Napier considered his patients' temperaments only very rarely, his medical judgments were primarily based on the two *causes* he drew from the astrological chart. These comprised a formal language with a finite number of possible "formulas" that specified in turn a closed set of humoral conditions.[197] This allowed him to work with a rather limited array of discrete medical problems that dictated treatment in a very straightforward way. It seems therefore that the *cause* in this particular meaning (i.e., the patient's humoral condition, which in Napier's practice was symbolized by a couple of astrological "formulas") played a similar role in practical astrological medicine to that of our modern *disease* today: it was the ultimate result of the process of medical judgment; it could constitute a coherent and finite taxonomy of medical problems; and it was the basis for consequent treatment.

If we understand *cause* in terms of the constitution of humors and regard *disease* (as a recurring pattern of symptoms) as only a secondary means of expressing medical problems, then we will not need to attribute to past afflictions either flexibility or contingency.[198] *Cause* and *disease* differed in many respects: *disease* was flexible in its definition, while *cause* was not; one *disease* could easily become another when the symptoms were reorganized in a new pattern, while a *cause* could transform into another only through a continuous (and usually mechanical) process; and while a *disease* was visible and superficial, the *cause* was hidden and essential. We can now better understand Forman's distinction between the two terms: "The sickness is on[e] thing and the cause is another, for the sickness may be the pox, the plague or the pills [piles], but the cause must be of coller, of fle[g]m, or of melancholy or some infections, humours or ayer [air]. . . . And I say except thou knowe the cause of the disease, and take it away, there can be no effecte. [I]t is not so much materiall to know the disease as to know the cause of the disease."[199]

Thus, the *operational identity* of disease (to use Cunningham's term), as it emerges from Napier's practice, was a firmly defined *astrological cause* that reflected the constitution of humors within the patient's body. This new outlook on the nosology of early modern disease, limited as it is to the context of astrological medicine, sheds, I believe, much light on the practical priorities of contemporary physicians and on early modern perceptions of illness and health.

The Course of Cure

The Logic of Therapeutics

Napier's patients traveled to Great Linford, sometimes from great distances, in order to find out what caused them pain or suffering; but more importantly, they came to be cured. Therapy was the doctor's most significant contribution within the medical encounter, and it was here that practical knowledge was finally turned into action. Not all of Napier's medical notes record a treatment plan, and the many personal prescriptions scattered throughout some of his manuscripts imply that he sometimes wrote the recommendations on a separate piece of paper. It is also possible that at some periods he kept a separate prescription book that was subsequently used by one of his assistants, who prepared and handed out the medicines.[200] Nevertheless, many of Napier's case notes contain specific orders for treatment, usually in the form of prescribed herbals and mineral drugs, less often as instructions on letting blood or the wearing of sigils. This means that a significant part of his records contains an overall account of the medical consultation, including complaints and symptoms, observations and judgment, as well as a recommendation for treatment, allowing us to reconstruct, at least to a certain extent, his underlying decision-making process.

As I have shown, the position of significant celestial forces, especially those that determined the *astrological cause*, revealed the condition of the humors inside the body, which itself specified the patient's affliction. Armed with this knowledge, the physician could then decide on the appropriate course of cure and choose his specific medicines. Since the medical problem lay in the excess of humors or in their corrupted mixture, practical humoral medicine mainly involved the act of purgation. This kind of therapy was widely employed by learned physicians as well as by popular healers and astrologer-physicians.[201] The choice of medicines was originally based on "cure by contraries": finding medicines with qualities that would counterbalance the patient's superfluous humor or initiate its evacuation.[202] Most astrologer-physicians, however, did not have to master the properties of plants and rather consulted pre-prepared lists of herbs and compounds that were associated with each humoral condition. Thus, the first part of Saunders's *The Astrological Judgment and Practice of Physick* described the humoral condition revealed by each possible *astrological cause*, while the second part listed the remedies that could manipulate and purge each of these conditions.[203]

The logic of therapy was fairly straightforward: once the malevolent humors were identified, they had to be evacuated from the body, either by acting on it externally or by invoking bodily discharge (purge or vomit) using internal drugs. Applying external procedures, most physicians combined outward remedies (such as unguents and plasters) with bloodletting or provoking sweat, as the former were specifically designed to draw the impurities to the surface of the body (the skin), from where they could be easily expelled.[204] Using internal medicines was somewhat more complicated. First, in choosing the appropriate purgatives the physician had to take into account not only the abounding humor, but also the parts of the body afflicted.[205] He then had to decide how to invoke the evacuation and through which orifice. "First consider," advised the seventeenth-century herbalist and astrologer-physician Nicholas Culpeper, "what the matter offending is, what part of the body is afflicted by it, and which is the best way to bring it out."[206] Basically, purgatives could either draw the corrupted or abundant humor to them, far from the infected organ, or drive it away.[207] Hence the significance that was attributed to the place of affliction. In addition, the physician had to decide whether to force the corrupted matter upward (by vomit) or downward (by stool). His task was not simply to move the bad humors away from their present location, but also to drive them in the *right* direction.[208] Astrological medicine included specific instructions for these issues that relied on various considerations, mainly involving the timing of the action, in ways that I discuss below.

Cleansing the body of the abundant or corrupted humors involved more than just provoking the desired kind of evacuation. The curative procedure consisted in fact of three separate steps and included medicines that were either digestives, preparatives, or purgatives.[209] Digestives were medicines that could cleanse the body of the malevolent humors without invoking evacuation. As their name implies, they digested the bad substance without moving it. Because of this, they were generally considered safer but less effective. Preparatives were designed to prepare the humors by making them motile and fluid before they were evacuated, and had to be taken for several days before starting the purgatives. According to William Salmon, a medical writer of the later seventeenth century, purgatives were to be given only after the body had been prepared in the following ways: "first, by exciting natural heat, and strengthening it; secondly, by sitting the humours, as if tough and 'clammy,' to prepare them with cutting Syrups; if thick, to attenuate them; and if the bowels be stopt or bound, to open them with an emollient Clyster."[210] Saunders explained, for example, that "*Acetosum flegma* must first be prepared well, for without preparation it will hardly be expelled, because it is slimy, thick and tough."[211]

Nevertheless, preparatives were not always needed: "you may purge without Preparatives if ♄ be in ♈," advised Saunders, "because the Humours be thin, but ♄ in ♌ requires Preparatives, because the Humours are more condense, and thicker, and causeth yellow thick Choler."[212] Here again, the "thickness" of the humor dictated the need for preparation. In other cases, digestives and preparatives could make purgation completely unnecessary. Robert Burton, author of the famous *The Anatomy of Melancholy*, advised that in the course of treatment "purges come last . . . [and] must not be used at all, if the malady may be otherwise helped." He also warned that if purgatives were to be used, it was better to begin with such that were gentler.[213] Napier does not seem to have taken the latter advice, as some of his most favorite remedies, specifically *diaphœnicon*, *hiera logadii*, and *confectio hamech*, were very violent purgatives. In Forman's case notes, preparatives and purgatives are often listed separately for a single case.[214] Napier, on the other hand, rarely made this separation explicit and normally supplied a unified list of drugs. It might be that he gave further instructions orally, or that such directions were given by an assistant who handed out the medicines. In one extraordinary example, Napier prescribed, "prepare 3 days beginning on Monday next, then purge, lastly use a lohoch."[215]

Reconstructing Napier's therapeutic decisions from his case notes is extremely difficult. The inherent complexity of the diagnostic process, together with the enormous variety of medicines that were attributed to each medical situation, makes it almost impossible to understand his choice of remedies even in cases that explicitly specify the *astrological cause*. Still, a careful consideration of specific cases treated by Napier reveals how the identified condition of humors led to a choice of particular medicines. For instance, in January 1603 Napier treated a two-year-old girl from Brafield who "eats nothing but drinks." He recorded the case next to the already drawn chart of the previous patient, and explicitly noted that her disease was caused "of ♀ in ♓."[216] According to Saunders's astrological manual, "Venus in Pisces" sometimes signified "venomous Humours and corruption of Blood, by venomous water and rotten blood." Saunders refers us to an adjacent discussion of "Venus in Cancer," where he advises to "let the Party sweat well, and then purge Flegm, and vomit."[217] Napier gave the sick girl *aqua theriacalis*, which according to Culpeper "expels venomous humours by sweat,"[218] and *syrupus de absinthio* and *aqua mentha*—both listed by Saunders as "Digestives of Flegm and cold Humours caused of ♀ in ♋."[219] To Mr. Underhill, whose disease was caused "of ♄ in ♏," Napier gave *diaphænicon, confectio hamech, syrup buglosse, aqua boraginis*, and *olei anisi*, which were to be boiled in a *common decoction*. Saunders's lists of digestives and purgatives for this humoral condition are rather long, but both *buglosse* and *boraginis*, as well as *confectio hamech*, are included in it.[220] *Diaphænicon*, on the other hand, was an electuary for discharging phlegm and an efficient lenitive. Napier prescribed this drug here either because, as he noted, the patient displayed an "abundance of flegme" or because he was "much grievous, sighing with his great pain."[221] Some medicines were prescribed for specific medical conditions that were basically humoral, but were identified and named as distinct medical conditions. For instance, when Forman discusses phlegm, he lists *diacuminum* and *diatrion piperion* among the remedies that are good for heating a cold stomach. Napier gave these two medicines to Mr Barton of Eastcote, who had "a very weake stomacke, cannot digest . . . stomacke wyndy." About two weeks later, when the patient returned with the same symptoms, he received another dose of *diatrion piperion*.[222]

The fact that, over and over again, specific medicines appear in a series of adjacent cases only to disappear completely when the state of the heavens slightly changes is another indicator of the asserted correlation between the patient's humoral condition, revealed by the state of the heavens, and Napier's choice of drugs. Still, two patients who came at exactly the same time did not always receive the same treatment. For instance, on January 15, 1603, a two-year-old boy named William was brought to Napier, who noted that the boy had a bad stomach and could not sleep. He gave him *unguentum rosatum* and prescribed *syrupus de absinthio* and *aqua mentha*, which were to be mixed in a *common decoction*. Next he saw Alice Rogers, a married but childless forty-year-old woman from Thornborough. Napier specifically noted that she came at the same time as the two-year-old boy and did not erect a separate chart for her. He diagnosed *tonsillarum inflammationem* (tonsil inflammation) and gave her *hyssop* and *diaphænicon*, the former being "good to wash inflammations" and the latter a strong pain soother.[223] Quite often, when the patient had very significant symptoms for which specific medicines existed, Napier treated the apparent symptoms rather than the hidden disease cause. For instance, patients who had difficulty sleeping were often treated with *syrupus de papvere erratico* or with *aqua lactuca*, the ingredients of both of which were known to induce sleep. Another medicine given in such cases was *diascordium*, an electuary[224] that was also known to have the same effect.[225] Other medicines, such as *electuarium contra tussim* (electuary against the cough) and *electuarium pro renibus* (electuary for the flux of the kidneys), were compounded specifically to combat certain symptomatic conditions.[226]

Timing Instructions

Choosing the appropriate moment for preparing and taking the prescribed medicaments was especially important in astrological medicine, and astrological guides were very detailed about the suitable times for the collection of medicinal herbs, their preparation, and administration. "If thou compoundest thy Medicine by chance, without election of a proper time," Saunders advised his readers, you will "affect the Patients Body contrary to thy intention, by virtue of Retention.... So be sure you make your Confection or Medicine in its proper Sympathetick Constellation, and then it must needs work Naturally, because of the Virtue that it receiveth in the time of its making by the Influence of the Heavens."[227]

In the early days of his career, at least, Napier bought his medicines from a London apothecary, so he probably did not have to worry about times of collection and preparation.[228] He did have to consider, however, the appropriate times for the administration of his medicines, since using the right technique at the wrong moment could easily result in no effect at all. Even worse, it could cause the corrupted matter to be pushed to inappropriate places, thus complicating the situation rather than solving it.[229] Thomas Bretnor and John Gadbury, two medical astrologers who were Napier's contemporaries, went so far as to declare that giving the correct medicine at an inappropriate time might result in the patient's death.[230] In fact, Napier himself warned in his "Treatise touching the Defence of Astrology" (hereafter "Defence of Astrology") that "a Physitian may administer a medicine in one constellatcion which may mortifie and kill the sicke, which ministered in a fitter constellation may either preserve him from death, or at least wise may procure him much ease."[231] This was based on both the theory of "critical days" in disease and on the notion that the direction of bodily fluids was governed by the stars and should be taken into account when choosing an effective therapy (especially purge versus vomit). At other times, the cure might just not work at all: "[If] The ☽ be in ♎, applying to ♀ in ♑ ... give no Purge to purge Melancholy or Choler, for it will not work ... for I have given to many at that time, and they never had above one stool."[232] Also: "Let not ♃ at any time be Lord of the Ascendant in the administering of a Purge ... for all this destroys the working of a Purge, and causeth it not to work, or to come upward, and so to be of no effect."[233] This also suggests that timing was significant in terms of the orifice that was chosen for the evacuation. As Forman explains,

> Vomitt: From the 20 of Marche unto the 20 of September it is good to purge by vomitt the upper parts of the body, because then the ☉, being in the north syde of the equinoctiall, also doth drawe upwards the humors of the body. . . . Esp[ecially] if ♃ be in [the] asc[endant] or be d[ominus] asc[endant] and the ☽ ap[plies] unto him or to any retr[ograde] pl[anet] in [the] asc[endant] or in 4 [the fourth house]. For if a man give or take any medicine almost inward to purge when ☽ doth apply to ♃ he will vomit it up.[234]

Celestial influence on the human body is revealed here in its most straightforward manner: the force of the Sun when it is over the Northern Hemisphere directly results in an upward draw of the humors. Napier himself seldom referred to rules concerning the administration of drugs, yet he sometimes gave specific instructions to start the treatment at some later point in time.[235]

A Multitude of Remedies

Perhaps the most striking characteristic of Napier's medical practice is his vast arsenal of herbal remedies. Leafing through the folios of his medical diary, one encounters a new medicine on almost every page. This is not entirely surprising, as it is already known that the early modern medicine cabinet contained an enormous and diverse collection of available remedies that could be exploited rather flexibly in medical treatment.[236] This confusing profusion did not escape the eye of contemporary writers on medicine, as in Burton's complaint about the treatments for melancholy: "I find a vast chaos of medicines, a confusion of receipts and magistrals, amongst writers, appropriated to this disease."[237] As I have already pointed out, Napier used most medicines in proximity, which meant that they were prescribed for a group of adjacent cases and then abandoned for a long period of time. Still, basic preparations such as the *common decoction*, Napier's favorites, such as *hiera logadii* or *confectio hamech*, and pain soothers such as *diaphœnicon* appear fairly regularly throughout his early career. The flexibility of choice embedded in medico-astrological rules was undoubtedly large. Forman ascribed dozens of drugs to each humoral condition, as did other guidebooks on astrological medicine.[238] Because each stellar constellation defined two *astrological causes* rather than one, the range of appropriate remedies was immediately doubled. In addition, each herb was linked to specific body parts and to one of the planets, based on its degree of heat and moisture, thus adding more options of treatment.[239] Sometimes this meant that different courses of therapy could be chosen for very similar cases. At other times it just meant that a single case included a very long list of remedies.[240]

By the first half of the seventeenth century, mineral, metallic, and alchemical preparations were already widely employed in English medical recipe books and printed pharmacopeias, and contemporary physicians were using them alongside their more conservative treatments.[241] Napier regularly prescribed his patients such cures, although more frequently in the later period of his practice. The only metallic medicine that he used during the early years of his career was *tabella stibium* (black antimony), of which Burton wrote, "Antimony or stibium, which our chemists so much magnify, is either taken in substance or infusion, &c., and frequently prescribed in this disease [melancholy]. 'It helps all infirmities,' saith Matthiolus, 'which proceed from black choler, falling sickness, and hypochondriacal passions.'"[242] Since Napier did not produce any written guides or commentaries on medicine or on drugs, his response to the religious and medical reform agenda of Paracelsus remains unclear. Largely led by a practical approach, adjoined to an inherent awe for certified ancient knowledge, Napier, like many other English writers and practitioners, it is probably fair to assume, absorbed the practical implications of Paracelsianism but ignored the rest.[243]

Outcome and Follow-Up

Napier occasionally revised his original interpretation of the chart or appended additional information to his notes. In one case, he noted retroactively that the patient "fell into a consumption of the lungs."[244] In another example, he went back to remark that the patient, a twenty-four-year-old woman from Harleston, "dyed about xi weeks after."[245] When a young woman from Deanshanger came to his consulting room complaining of weakness, fainting, vomiting, severe pains, and stopped menses, Napier did not record any medical judgment, nor did he prescribe any remedies. Several weeks later, he modified his original notes, noting a third aspect to the Moon and adding under the chart, "proved [to be with] a child & dyed in child bed & her child also. tooke much physick to destroy the fruit of her wombe. delivered of a very poor child dead."[246] Referring back from the current visit to an earlier consultation of the same patient was also a common routine, and Napier often noted the date of the former encounter under the astrological chart, revealing that he practiced some kind of medical follow-up. Kassell suggests that astrologers noted the times of former visits because the moment in which the first question was posed was crucial for later calculations. However, Napier's references usually point to the patient's *previous* visit rather than to the first. This lends support to the claim that his main concern was linking the current visit to the context of disease narrative so as to maintain some degree of treatment continuity.[247]

Continuity of treatment is no doubt one of the most important features of modern medicine. Within a series of related medical visits and until our problem is solved we expect our doctor to stick to some kind of "course of cure." Of course, he or she can choose another route of therapy or simply change their minds along the way, but we assume that each visit will be based on formerly acquired knowledge and on the results of previous treatments. The treatment we receive at each visit should have a positive impact on our symptoms or a measurable negative effect on what we see as the "disease essence." If it does not, or if the impact is not satisfactory, we expect the treatment plan to be changed. Our condition is repeatedly assessed either by examining us physically or through additional laboratory or auxiliary tests, the results of which need to be carefully evaluated as a token of our current medical state. In Napier's consulting room, the success or failure of the treatment was measured in a much more straightforward way. Since most of his medicines were purgatives or vomitories, their outcome was visible and measurable, simply by counting stools and vomits. Napier did not just check his previous notes to be reminded of the original problem and compare the patient's former state to his current condition.[248] He was also interested in the outcome of the purgative treatment he had prescribed, according to which he adjusted or complemented his course of treatment. For example, on October 14, 1602, the thirty-eight-year-old Thomas Johnson of Crendon complained of a great ache in his left hip, "which runnes downe to his foote [on] the outsyde of his legge." Napier proclaimed that the pains were caused by sitting on a cold stone, and gave him two pills of *fœtidæ*. A week later, when the patient returned, he noted that the prescribed drugs were taken as ordered and that "they wrought not untill yesterday, then he had 5 stooles, but his ache in his hippe continueth." This time

Napier prescribed a special plaster made of rye bread, calf meat, egg, and saffron, hoping to defeat the disease by external means.[249] Of another patient, Napier noted that "the losenges [troches] gave her two vomits, the purge 3 stooles, the payne in her side gone, willing to be let blood."[250]

Counting vomits and stools was not exclusive to Napier and appears in the case records of other contemporary practitioners. John Hall, an English physician and Napier's contemporary, made known the results of his treatment by stating the number of stools or vomits they produced. Of one of his patients, a fourteen-year-old boy, he reported,

> *Richard Wilmore* of Norton, aged 14, vomited black worms, about an inch and half long, with six feet, and little red heads. . . . I gave the following remedies: ℞ *Merc. Vitæ* gr.iii. *Conserv. Ros. Parum*. This gave seven Vomits, and brought away six Worms. . . . The following day I gave this: ℞ *the Emetick Infusion* ʒv. It gave five Vomits, and brought up three Worms. The third day I gave the following: ℞ *Spec. Diaturb*. . . . This purged well, but brought away no worms. Thus he was delivered, and gave me many thanks.[251]

It seems, then, that at least some seventeenth-century physicians, operating within the humoral framework, exhibited an early modern version of follow-up and continuity of care: two aspects of Hippocratic-Galenic medicine that were somewhat overlooked in previous historical research.

Astrology, Medicine, and Knowledge

My painstaking investigation into Napier's practice reveals how his techniques and procedures led him to conclusions and actions that were very similar to those of contemporary nonastrological physicians. Like them, he mainly dealt with the evacuation of corrupted or superfluous humors, but in order to reach his conclusions he employed judicial astrology as a vital tool of his art. Although he only rarely used it for nonmedical purposes,[252] or to predict the outcome of an illness, its critical days or points of crisis,[253] astrology played a central role in Napier's assessment and interpretation of medical conditions. Astrologer-physicians such as Napier were not quacks. In fact, Napier took his art very seriously: meticulously following its strict procedures, correcting findings retroactively in order to improve his skills, and often practicing it on himself. For him astrology was not some supplementary technique, another tool in his professional toolbox, but rather the core of his professional expertise.

In his "Defence of Astrology,"[254] written on request as a preface to Forman's "Astrologicalle Judgmentes of Phisick and Other Questions,"[255] Napier declared that knowledge of the heavens is "soe necessarielie required to a Physitian, that hee cannott bee . . . reckoned an absolute & skilfull Physitian, that is utterlie ignorant of this Astrologie."[256] "[T]here is as much difference to be putt betwixt Common and vulgar physicke done without any regard had of Astrology, and Physicke Joyned with Astrology," he quoted Ficino, "as is betwixt wine and water."[257] What made judicial astrology so indispensable to Napier's practice? My answer in the following is predominantly functional. It applies to the particular kind of astrology that was practiced by Napier; for purposes of convenience and accuracy, I will refer to it here as "Forman-Napierian astrological medicine." More generally, I will argue that astrology and an astrological worldview were central to Napier's philosophical and theological outlook, and show how he maintained their legitimacy and truthfulness using the notions of "cumulative knowledge" and "general consent," which gave grounds to his overall view of true and certain knowledge.[258]

Historians who have dealt with the popularity of astrology in seventeenth-century medicine have stressed its advantages as both a theoretical and a practical system. Keith Thomas argued that the primary contributions of astrologers lay in their ability to provide information that other professionals could not and in offering advice to those in a state of helplessness. He also suggested that astrology offered a kind of compromise between harsh determinism and the capricious free choice of men, because it revealed the reasons for clients' misfortunes while, at the same time, giving them sufficient information to take responsibility for their own destiny.[259] Others have stressed its role in deciphering the relationship between the different planes of existence: corporeal, cosmological, and social, thereby creating meaning and allowing patients to understand why they were sick.[260] The procedural role of astrology has also been acknowledged, with attention drawn to the way in which the astrological chart imposed a strict format on consultations and records.[261] These conclusions are very important to the understanding of the astrological-medical framework that guided Napier and his contemporaries. However, in the main they leave out its significance as a technique that had considerable practical advantages for the early modern physician.

In early modern medicine the act of judgment was by far the physician's most difficult task. The ambiguity and complexity of the evidence, the need to pick out the signs of disease from among numerous physical symptoms while considering at the same time the patient's individual complexion, entailed a very intricate procedure. Not only was the number of possible signs vast, but different signs meant different things as well: "There are certain signs that will always be present whenever a catarrh is involved; other signs will reveal the particular pathway or pathways along which the flux is occurring; others will identify the qualities of the humour involved; and others will determine the ultimate source of the humour."[262] This left a great deal to the doctor's judgment, to the point that he sometimes preferred to be tentative rather than decisive.[263] Moreover, a symptom could only be a sign if it was not within the norm for the specific patient, but no standards existed to allow early modern physicians to compare individual symptoms with some "normalized baseline." Forman-Napierian astrological medicine, on the other hand, was based on a finite number of "formulas" that indicated in a fairly straightforward manner the patient's underlying medical condition. Supplying the necessary "hard evidence" for making a medical judgment, it served as an excellent tool for figuring out the "matters of fact" in Napier's consulting room. It also provided information that was regarded as sufficiently grounded and practically indisputable by both doctor and patient, allowing Napier to be efficient and practical and at the same time to position himself as an authoritative professional and present a reliable disease narrative.

Keith Thomas maintained that seventeenth-century astrology, while claiming to be an objective science, was in reality highly flexible, since it left room for infinite possibilities of disagreement over the interpretation of the chart. The number of possible combinations and permutations open to the physician was so immense, he argued, that "to pick his way through them he needed not mere technical skill, but judgment." Any interpretation of the astrological data was therefore bound to be subjective.[264] However, such a view refers to astrology as a whole rather than to a specific method practiced by a particular practitioner at a particular period of time. Although the number of parameters in the astrological chart was indeed vast and could be interpreted in numerous alternative ways, the interpretation of a specific astrologer who followed a certain set of rules was much more limited. As I have shown, Forman-Napierian astrological medicine was based on two major *astrological causes* that specified the patient's medical condition from within a closed set of humoral states. This limited "formal language" supplied Napier with a simple, fast, and unambiguous way to reach a medical conclusion—certainly when compared to the evasive method of symptoms and signs that traditional Galenists had to deal with.

Medical diagnosis has always been based at least to some extent on some "hard evidence": a grain of certain truth that anchors an otherwise uncertain process.[265] This grain plays a crucial role in convincing the patient that the doctor's interpretation of his own subjective experience is trustworthy. A modern physician might point at returned lab results and say something like "Look! Your blood glucose is over 170, no wonder you are constantly tired and thirsty."

The sheet of lab results she is holding in her hand is the "hard evidence" that underlies what will shortly be her diagnosis; in this case, *diabetes mellitus*. The majority of medical practitioners in the early modern period, both learned and popular, lacked such reliable sources of information and an appropriate framework for their meaningful organization. Instead of a complex interpretation based on mostly ambiguous symptoms, astrology offered an external source of knowledge that was assessed in a rather straightforward way but, at the same time, mirrored the patient's underlying medical condition. Its signs and states were visible and comprehensible (at least to the professional eye) and were therefore seen as observational givens. With its reflection of inner bodily processes, its visual method, and its formal symbolic language, astrology supplied Napier his needed "hard evidence." Like our modern doctor, he too could point at the chart, look back at his patient, and confidently state his medical verdict.

Such conclusive medical judgments, referring as they do to matters of health and touching at times on matters of life and death, have to be received as authoritative. In order to establish such authority, early modern physicians had to gain credibility for their methods, but also to provide disease narratives that would be accepted as "matters of fact."[266] Andrew Wear has pointed out the tendency of these physicians, not only to frame disease in narrative form, but also to attempt to "prove" their stories, that is, to support them with observable proof.[267] This urge to provide "hard evidence" also guided astrologer-physicians like Forman and Napier. Forman used his examples as an opportunity to establish the truthfulness of his methods, connecting his disease stories to the direct evidence that he found in his patients' humoral discharge:

> If ♄ be *d[ominus]* 6 in ♍ *post plenam lunam* and in 12 which is his house of joy, and ☽ do sep[arate] from ♄ and ap[ply] to ♃ in ♈, ♃ being *d[ominus] h[ora]* in 8, heere ♃ hath most dominion and sheweth the melancholy humor hath overcome the bloud altogither, and that there is some yellow chollericke water mixed with it. Experience by Leonard Shawe, for he vomited much choler, and when I opened the liver veyne in the left arme, that which came from him was all melancholy and blacke dreggs mixed with yellow water and it was rotten and feculent. And the like I did 7 tymes after againe, and found it even so againe, and but litle altered till the third tyme. The like againe was seene by Ellen Kellor 1596 March, where ☽ did ap[ply] to ♄ *d[ominus]* 12 in *novi lunio*, ♄ in ♍ ret[rograde] in 6, and ☽ did sep[arate] from ♀ in ♒ in 12, ♃ being *d[ominus] h[ora]* in ♈ at a ✶ to ☿ in ♑, where the disease was caused of much melancholy mixed with choller, but melancholy had the dominion, and her bloud was

altogither rotten and putrified, and I let her bloud twise and found it to [be] so.[268]

Of course, this was essentially similar to the kind of evidence presented by physicians who practiced nonastrological medicine, since it validated the humoral judgment against the direct observations of the patient's excretions. However, it does mean that astrology played a significant role in building the authority of astrologer-physicians and that establishing authority was important for their ability to practice medicine. The patient's acceptance of (if not belief in) astrological-physic played an essential role in its success as an alternative to traditional methods of medical judgment.

Yet astrology had something more to offer that could enhance the validity of its disease stories: it supplied a narrative that portrayed change over time. What Napier and his fellow astrologers saw in the chart was not a snapshot of the body's current state, but rather the captured motion of its alterations over time. The planets in the chart were not simply "there," in this or that house; they were always moving in some direction: applying, separating, retrograding, and so forth. Their exact place at the time of the question uncovered the forces that ruled the body at that particular moment, but also revealed what it had gone through and what was going to happen in the future. Medico-astrological rules often contained disease stories that described the course of disease and the inner paths of pain. This matched the very similar stories of patients who often complained, as we have seen, of "wandering pains." Astrological medicine supplied, therefore, not only a glimpse of the hidden cause of disease, but also a consistent method for reading a narrative that paralleled that of contemporary patients. The ability of astrological symbolism to capture and portray such narratives played a major role in establishing its authoritative status as a vehicle of true judgment.

This is still, however, only part of the story. Fitting as it did within a broader astrological framework, Napier's adherence to practical astrology was in fact integral to his overall philosophical and theological outlook. As one of several channels of divine wisdom, astrology constituted a linguistic key to unlocking nature's secrets and a vital tool in Napier's active search for true knowledge. This is perhaps why, despite some initial reluctance, he accepted Forman's challenge to utilize his scholarly talents (which Forman apparently lacked) to shield their mutual art against critics and adversaries.[269] The resulting "Defence of Astrology" provides an excellent illustration of how Napier generally conceived of the establishment and confirmation of trustworthy knowledge.[270] Particularly important for my argument are the notions of "cumulative wisdom" and "general consent," reiterated and extended in his later letters, apologies, and theological tracts. Employing these two concepts, Napier maintained that the legitimacy and credibility of astrology draw on an unbreakable chain of divinely transmitted knowledge, passed on through a few ancient adepts to scholars of all ensuing generations; and from the consensus thus created,

> I doubt not but this Art of Judiciall Astrology will also deserve (if Credit may be given to aunciient storyes & records) not only her love and Admiration, but also her praise and due Comendation.... we fynde it recorded by good and substatiall authors that the very first inventers therof were the holy patriaks [patriarchs] who, receiving that knowledge from God himself, delivered the same touching the necessary use therof to their Children, and so it being by a Certyne traditionary succession Derived from one to another, hath ever since through the gratious and mercifull providence of the Almighty Continued ... even to this age of ours.[271]

The authenticity of astrological medicine, claimed Napier, builds first on the divine origins of astrology itself, then on its place in the teachings of ancient scholars and past men of fame, then on its employment by numerous medievals and moderns, and finally on the daily experience of men like himself:[272] "sithence this opinion is stronglie confirmed by scripture, authorities of Philosophers, Astrologers, Devines old & new, approv[ed] by dailie experience, & observacion of the celestiall bodies, & by the wittnesse of holie auncient fathers, & by the consent of all nations, & people."[273] "[A]ll these affirme & avouch with one generall consent," Napier pronounced, "that the heavenlie bodies make their action & impression upon these earthlie bodies, which lye here beneath them."[274] The special authority of cumulative knowledge draws therefore on the "general consent" it creates across generations.

Moreover, each individual link in the chain of transmitted knowledge itself rests on widely held agreements, rechecked and reproved repeatedly against contemporary experiences and the assertions of various scholars. Napier's "general consent" is therefore twofold: referencing both the joint consent of all people at a particular period and that of all generations. Thus, the French theologian Lambert Daneau had proved his opinion by the "*Omnium gentium et popularum consensus,* [i.e.,] by the generall, & uniforme consent of all natures, & peoples,"[275] as also "by the dailie & continuall experience of all ages."[276] In an early echo of his disapproval of the individual interpretation of the Bible, the lone dissenting voice of Napier's Flemish contemporary Andreas Gerhard Hyperius is rejected as the "private fancy" of a single man:[277] "Hyperius dennyeth this position of Aristotle as false and flatly repugnaunt to the scriptures. ... But this opinion of hyperius though never so learned, may in no Case be admitted or approved, because it is but one mans opinion and that founded but upon his oune private Conceat and fancy and different from all others, induced By weake arguments, reproved and Controlled by the generall Consent of the learned."[278]

Having established astrology's authenticity and reliability, Napier goes on to consider its lawfulness within the Christian framework. The employment of judicial astrology in practical medicine rested upon the alleged correlation between the instantaneous heavenly constellation and the concurrent state of humors inside the human body. Drawing on the notions of principal and secondary causes and of the cosmos as a preset clock, Napier contends that such a link hardly challenges God's perfect omnipotence: "[O]ur question is not ... what god might, or could doe, but what pleased him to doe, ... that god in his unsearcheable wisedome ... hath ordained that there should bee first & second causes, & that the world in his government should resemble a clocke in his motion, & god as the principall wheele drawing on the second, & they the rest, working as the first agent, & principall cause, using the heavens & the starres therein as a second cause in the moving, altering, & disposing of thinges beneath."[279] On the contrary, Napier argues, one who ignores astrology "[does] not greatlie regard the workes of the almightie & their necessarie uses."[280]

Through this apology for his practical art, Napier reveals the pillars of his philosophical outlook: an inherent awe for cumulative, consented knowledge constantly matched against the daily experience of learned men. Authenticated and authorized accordingly, astrology served as a symbolic framework that allowed able men to acquire certain practical knowledge concerning human health. As my next chapter will show, it could also be used as a magical language by means of which one could act proactively on the cosmos, manipulating its natural forces to achieve desired medical outcomes.

2

ASTRAL MAGIC

Napier's name as a magus arose from an early, if post-factum, reputation as a master of magical arts, cemented by a modern portrait of an "obscure and rustic Faustus."[1] Yet the extent to which the renowned doctor was practically involved in the various magical traditions of his era is not entirely clear, as the scale and diversity of the archives sometimes obscure the facts of his actual daily work. One magical tool that he did undoubtedly utilize was "astrological" (or "astral") magic: a set of techniques designed to exploit celestial influence through the employment of astrological talismans, worn or otherwise carried on the patient's body. An intriguing story concerning Napier's use of such an enchanted object to cure a sick maid appears in Lilly's autobiography:

> A maid was much afflicted with the falling-sickness, whose parents applied themselves unto him [Dr. Napier] for cure; he framed her a constellated ring, upon wearing wherof, she recovered perfectly; her parents acquainted some scrupulous divines with the cure of their daughter; the cure is done by inchantment, say they; cast away the ring; it's diabolical; God cannot bless you, if you do not cast the ring away. The ring was cast into the well, whereupon the maid became epileptic as formerly, and endured much misery for a long time. At last her parents cleansed the well, and recovered the ring again; the maid wore it, and her fits took her no more. In this condition she was one year or two; which the Puritan ministers there adjoining hearing, never left off, till they procured her parents to cast the ring quite away; which done, the fits returned in such violence, that they were enforced to apply to the doctor again, relating at large the whole story, humbly imploring his once more assistance; but he could not be procured to do any thing, only said, those who despised God's mercies, were not capable or worthy of enjoying them.[2]

Stories like this ensured that Napier's early name as a master of occult arts persisted for more than a century after his death. Such stories also fostered tensions between him and neighboring ecclesiastics, alarmed by some of the techniques employed in his thriving clinic. Still, as I will show, Napier's use of astrological talismans was not only an effectual but also a lawful part of his medical practice, resting upon the very foundations of his philosophical and theological worldview.

Sundry collateral or secondary evidence seems to associate Napier with further magical techniques, such as charms and exorcisms. One of these is a short piece of parchment containing a text of ritual dispossession, written for "Mr. E. Fr.," who was suffering from melancholy and was apparently paralyzed. Its author orders the "cruel beast with all the associates & all other malignant spirites" that have taken charge of the poor man's body "speedely to departe from this creature & servant of god . . . [and from] every parte of his bodie."[3] This short ceremonial script, the only one of its kind attributed to Napier in William Black's catalogue, is neither signed nor dated, and is evidently not written by the doctor himself. Its attribution to Napier is probably rooted in its inclusion in the same casebook as some of his letters, astrological observations, and memoranda.[4] In this case, therefore, the available evidence turns out to be rather elusive and certainly insufficient to ground Napier's participation in this kind of magical healing.

The use of charms and spells was also allegedly incorporated in Napier's medical practice. The English biographer and antiquary John Aubrey included the following story in his short biography of the late doctor:

> Mr. Ashmole told me, that a woman made use of a spell to cure an ague,[5] by the advice of Dr. Napier. A minister came to her and severely reprimanded her, for making use of a diabolical help, and told her, she was in danger of damnation for it, and commanded her to burn it. She did so, and her distemper returned severely; insomuch that she was importunate with the doctor to use the same again; She used it, and had ease. But the parson hearing of it, came to her again, and thundered Hell and damnation, and frightened her so, that she burnt it again. Whereupon she fell extremely ill, and would have had it a third time, but the doctor refused, saying, that she had contemned and slighted the power and goodness of the blessed spirits (or angels) and so she died.[6]

Apart from being conspicuously similar to the story quoted from Lilly (with which this chapter opened), this account is not supported by any evidence in the archives. Although the learned rector evidently valued the power of words, there is therefore hardly any indication that he ever used anything other than sanctioned forms of Christian prayer. Indeed, Napier apparently attempted to hide his involvement with any kind of magic, at least partly, either by separating any revealing written evidence from the main body of his manuscripts, or by avoiding such documentation altogether. Consequently, some of his magical material found its way into the Sloane Collection, together with similar texts by Simon Forman and Elias Ashmole. These too contain scanty evidence of Napier's actual involvement in any written or verbal magic. At the very best, they might indicate his scholarly acquaintance with such ideas and methods.[7] Napier's participation in another kind of magical activity, however, the invocation of angels, is evident from numerous remarks scattered throughout some of his medical diaries, as well as from a handful of folio papers recording his short queries to the archangel Raphael and his alleged responses. As I will show in the next chapter, these spiritual conversations supplied Napier with an especially reliable source of information on his clients' recovery prospects and their possible courses of cure, and on personal questions of longevity and prosperity, as well as more general insights into medicine, alchemy, and religion.

In the first chapter, I drew the line between astrology and medicine at the division between *knowing* and *acting*. Following the same divide, I intend to locate my discussion of Napier's astral magic within the domain of the latter, that is, within the operative aspect of his practical medicine. His conversations with angels, on the other hand, will be primarily dealt with in the context of knowledge acquisition.[8] My distinction, therefore, will be a functional one: this chapter will deal with Napier's magical remedies, while the investigation of his angelic conversations will be postponed to the next chapter, discussing divination.[9] In terms of introducing categories of magic, however, the present discussion serves as a preface to both.

I will therefore open this chapter with a review of the various medieval and early modern magical traditions, focusing mainly on the distinction between "ritual" magic and "image" magic and on the theoretical framework of astrological talismans. I will then portray the making and use of such talismans, "sigils" in Napier's terminology, as part of his daily medical practice. As I will show, techniques of astral magic became an integral part of Napier's medical treatment, adding further tools to his therapeutic arsenal.[10] In conclusion, I will consider the legitimacy of astral magic from Napier's point of view, building on the Scholastic categorization of magic and on his original apology for the use of carved images. As this chapter's opening citation suggests, criticism of magical activities was inherent to its practice. Yet the employment of astrological sigils posed no significant difficulty for Napier's theological or philosophical worldview.

Kinds of Magic

Magic and medicine were long intermixed in Western healing, making the distinction between naturalistic and magical therapy unclear as late as the early modern period. The medieval medical manuals known as "leechbooks," for example, which mainly contained recipes of herbal medicine, often prescribed magical charms and spells for healing as well. Even learned medieval physicians recorded magical remedies, and herbal medicines were "often administered, or recorded, in conjunction with charms."[11] Healers, divines, and parish priests who wanted to address their flock's inclusive needs were traditionally involved in an eclectic set of natural and magical practices that were aimed at healing both mind and body.[12] Their arsenal of "magical cures" included the using of amulets (natural objects allegedly capable of producing a positive effect), talismans (natural or artificial objects decorated with artificial marks, designed for the same purpose), written or oral charms, spells and exorcisms, and other techniques or rituals intended to invoke divine or supernatural help.

Only late in the medieval period did Scholastic rationality and the church's monopoly over the supernatural come to demand a clear intellectual division between different types of magic. A fundamental distinction was now made between so-called natural and demonic magic: terms that referred to the mechanism underlying the magical action rather than to any discernible features. Thomas Aquinas, one of the first Christian authorities to set some intelligible rules in this realm, differentiated in his *Summa Contra Gentiles* between various kinds of talismans according to the nature of the images they bore. He maintained that a talisman became sinful only when its figure included signs to be interpreted by a separate intelligence (i.e., words, letters, or other intelligible characters), and accepted a natural foundation for the power of talismans that did not include such signs or forms. This distinction, referred to by Brian Copenhaver as a division between "noetic" and "non-noetic" magic, and by Nicolas Weill-Parot as the question of "addressativity," introduced a new classification of magic into the Christian framework.[13] An "addressative" magical act, that is, one understood as aimed, either by the use of signs or by the use of rituals, at a supernatural being, and situated outside the divine order and sanction of the church, was now condemned as demonic, while "non-addressative" magic was accepted as potentially "natural" and lawful.

Modern historical research on medieval magic, however, tends to classify magical activity in terms of "ritual" magic and "image" magic, where the former is characterized by an emphasis on the ritual action as the basis of the magical result, and the latter by its working through the static power of signs or forms.[14] An important feature of ritual magic was that its operators understood it necessarily as being directed toward some sentient beings that could be conjured or summoned, either in a direct vision (whether in waking life or in a dream) or through a human medium. This often required the magus to achieve higher spiritual states, usually through a set of meditative exercises. Authors of ritual-magic texts, who could not allege its ascription to any explicable forces of nature, had to base their justifications in claims to divine authorization or inspiration. As a result, argues Frank Klaassen, "their chief 'claim' to legitimacy lay in their orthodox appearance, internal arguments which elaborated upon orthodox ideas and sources, strenuous adherence to traditional notions of piety in their rules and practices (abstinence, prayer, mortification, etc.), or claims to direct sanction by the divine."[15] As we shall see in Napier's case, image magic sometimes contained ritualistic components, making the modern distinction harder to apply. Still, I find its functional categorization more helpful for my purposes, as the alternative distinction, between "noetic" and "non-noetic" magic, would probably have been rejected by Napier, who saw all kinds of magical activity as directed solely at God's emanated attributes.

"Image magic" dealt with the deliberate drawing down of spiritual power or celestial virtue by largely static means, mostly figures or signs. Its techniques were subjected to two fundamental and related principles: (1) the correspondence of all parts of the operation, and (2) the influence or transfer of power from superior to inferior forms. The first principle required the images or forms to be drawn at precalculated moments, and on the "appropriate" material; the second was based on the assumption that certain objects bearing the images or signs of higher forms (either natural or supernatural) could be used to perform earthly actions associated with their specific influence.[16] The talisman, into which celestial power was usually transferred, was the most common tool in such magical operations. Talismans employed in image-magic operations were thus static tools of practice, although the process by which they were produced sometimes incorporated certain ritualistic actions such as incantation, fumigation, or the employment of special garments, flowers, or branches.[17]

Image-magic traditions, especially those related to astral magic, rested primarily on Arabic texts that were translated into Latin around the twelfth century,[18] and to a lesser extent on Greek and Hebraic sources. Among the prominent works in this domain are *Liber Lunae*, *The Picatrix*, and Thâbit ibn Qurra's *De Imaginibus*. Some of these texts are believed to originate from Persian and Indian sources, in particular from the Sabian community of star worshippers at Harrân.[19] Renaissance works of image magic, including those of Marsilio Ficino and Henry Cornelius Agrippa, gathered together much of the existing practical literature[20] but wove into it so-called hermetic ideas,[21] as well as Neoplatonic and kabbalistic material.[22] Ficino presented his image magic within the traditional medical framework, stressing its potential applications in this domain.

Forman donated his share to this written tradition with several original works. His major unpublished treatise, "Of appoticarie druges,"[23] was a comprehensive review of medicine, magic, astrology, and alchemy, containing entries on various kinds of magical operations. In his expositions "Of microcosmus or man,"[24] "The arte geomantica,"[25] and "The motion of heavens," he combined notions from Christian cosmology, Renaissance Neoplatonism, and Arabic astral magic.[26] Napier owned numerous books and manuscripts of magic, some of which are mentioned in his professional correspondence or named in the carefully recorded lists of books and manuscripts that he regularly lent to his colleagues and friends.[27] The most prominent of these, *The Picatrix*, probably came into Napier's possession through the acquisition of Forman's library after the latter's decease.[28] He also owned Agrippa's *De Occulta Philosophia* (perhaps the same copy that Forman had bought in 1580)[29] and Petrus Constantinius Albinus's *Magia Astrologica* (published 1611),[30] and was at least acquainted with Balamin's *De Sigillis Planetarum*[31] and Laevinus Leminius's *De Occultis Naturæ Miraculis*.[32]

Avoiding, as usual, anything lacking a practical aim, Napier refrained from making written contributions of his own. As in other fields of practice, the traces of his own genuine knowledge must be cautiously picked out from within the records of his medical work and from the theoretical writings of his tutor. Two primary sources of information on Napier's practical techniques are MSS Sloane 3822 and 3846, which include various occult materials by Forman, Ashmole, and (albeit more scarcely) by the doctor himself. The former includes a few dozen folios of calculations on the appropriate times for making sigils, and explanations concerning their production and use. Many of these notes are accompanied by illustrations of astrological talismans, which bear symbols of the planets, signs of the zodiac or their visual representations, or kabbalistic words. The scant materials in Napier's hand include instructions for the making, wearing, and use of specific sigils,[33] as well as a learned apology for the use of graven images, with which I will deal at length toward the end of this chapter. Sloane 3846 is mostly occupied with texts and rituals for the summoning of angels and spirits, but also with insightful instructions on the making and use of certain sigils. Also bound inside the same volume are copies of some major astrological image-magic works, such as *Liber Raziel, Liber Lunæ*, and *De Imaginibus*, neatly written in Gerence James's hand. As most of the examples I bring forward in this chapter are Forman's and Ashmole's, it should be kept in mind that linking these materials to Napier's daily practice is mostly conjectural. I allow myself a certain fluidity here, based on the reasonable assumption that

Forman's manuscripts, eventually owned by Napier, were used as guidelines in his practice, and that both took part in one and the same practical tradition, later studied and analyzed by Ashmole.

Apart from the scant but significant traces in his medical diaries, the only evidence of Napier's involvement in astrological magic is to be found in a few sporadic paragraphs in a small number of his letters. These small remnants of what seems to have been a diverse correspondence among practitioners of natural magic reveal some of the ways in which practical knowledge traveled betwixt Napier and his colleagues: a community of learned scholars subdivided into circles of professional interest. Some of these letters reveal how Napier himself was consulted by novice magi who already then considered him a skilled operator of astral forces. In a letter received in 1626 from his brother-in-law Sir Thomas Myddleton,[34] Napier is consulted about the making of a golden sigil, prepared according to his former instructions:

> [H]aving receiued your letter and box, I addressed my selfe to the effecting of itt and to that purpose went and gott a peece of gould beaten thin, and about halfe an hower before seven, gott the engraver to begin to worke vpon itt and to engrave on boath sids of the gould according to the patternes you sent, being boath one as I take itt. . . . If I have done right, I am glad, if nott, I desire to be informed from you, how to amend itt. I purpose to have a stampe made, and to have two or three peeces of gould reddy, and to stampe them att such tyme, as you shall finde fittest, of which tyme I shall desire you to informe me.[35]

Always willing to offer learned advice to a novice colleague, Napier became a contributor of magical knowledge despite his abstention from any scholarly writing. His astral magic lived on after him, studied in depth by younger admirers of the mid-seventeenth century such as Lilly and Ashmole.[36] For a brief moment in history, Napier was counted an esteemed heir to a long list of learned magi, promoting the magical art of healing through the manipulation of celestial forces.

The Nature of Astral Magic

Astrological Sigils and Celestial Forces

Reluctant as he was to share his secrets with others,[37] Forman was apparently prepared to make an exception when it came to his diligent Great Linford apprentice—a generosity not only highly appreciated, but also adequately rewarded. "Youre louinge and kind letters," he wrote to Napier a couple of months before his death in 1611, "came vnto me by Rutland the carrier, who brought withall 2 cheases and a lyttle booke of Merline, as tokens of remembrance of youre kinde good wills [and] carfull courtasy."[38] Contented with the offerings he received, Forman replied with a set of brazen molds designed for the casting of astrological sigils, coupled with the necessary instructions concerning their making and use:

> Touchinge those brason mouldes for caractes [characters] of the plannetes, yf youe haue them, and can tell howe to vse them, youe haue a good thinge aswelle for the cueringe of diseases as for diuers other purposes, to caste therin in mettalle the sigelle of any plannet when he is stronge in the heauens, for the effectinge of any purpose or thinge pertayning to that plannet or when ther is a conjunction △ or ✶ of 2 plannets with reception, to caste the caractes [characters] of the sam[e] plannets at that instante, makinge the on[e] Lord of the ascendente & the other Lord of the howar [hour].... [A]nd soe shall they be of gret effecte and caste them alwais in that mettalls that belonge vnto the plannet.[39]

Forman believed that such objects, when properly made, "enclosed som[e] parte of the vertue of heaven and of the plannets,"[40] and prescribed sigils, laminas, and constellated rings to many of his patients, alongside his more popular treatments of purging and bloodletting.[41] As with much of his practical art, Napier learned the secrets of astrological talismans, including their uses, making, and application, from Forman, the self-made magus.

The technique of using astrological talismans, known as astral magic (Forman referred to it as "astromagic"), was based on the idea that occult powers, emanating from the planets and stars, could be channeled at astrologically propitious moments into man-made objects, sometimes through the agency of named spirits and angels. These hoarded powers could then be used, even months or years after the talisman was "charged," to bestow a positive or negative effect on the lives and bodies of human beings. Celestial forces, invested in astrological talismans, operated through the same principles of sympathy and antipathy that were utilized in astrological herbal medicine. For instance, since the Sun was astrologically associated with gold, a golden talisman engraved with the image or symbol of the Sun, at the moment of its strongest influence, was believed, through antipathy, to protect against lunar diseases such as smallpox.[42]

The notion of occult qualities and their influence on the human body was already familiar to ancient physic. Even Galen, who preferred to explain the action of foods and drugs on the body in terms of their manifest properties (hot, cold, dry, and wet), was sometimes forced to resort to explanations based on so-called indescribable properties (*idiotetes arretoi*). In fact, many substances in Galen's pharmacological works, including certain foods, drugs, poisons, and amulets, were said to work by a "certain power of attraction," or partly "because of a property of the whole substance, . . . [partly] because of their active qualities."[43] Avicenna, probably the most prominent contributor to the Galenic tradition, explained that certain substances act on the body, not because of their matter or their manifest qualities, but through their specific form.[44] These two concepts of "whole substance" and "specific form" would subsequently play a major role in the Thomist theoretical framework for astral magic.

The translation of Arabic texts in the twelfth and thirteenth centuries, building on the same ancient tradition, vastly increased Western knowledge of occult powers and influenced medieval philosophical and Christian writings such as Thomas Aquinas's *De Occultis Operibus Naturae* (On the Occult Works of Nature). Thomas's use of astrology's underlying principles in explaining certain mysterious phenomena in nature prompted, and in effect directed, much of the later Christian discussion on astral magic.[45] Renaissance Neoplatonism developed the notion of occult qualities and influences into a full-fledged theory of the cosmos as a living system, guided by a "world soul" or "cosmic spirit."[46] Ficino, perhaps the most prominent author on Christian magic in the Renaissance, not only stated in the abstract that the macrocosm constantly affected terrestrial objects, but also isolated and analyzed its specific influences. In his *De Vita Libri Tres* (Three Books of Life), he explained how these occult properties worked:

> At the same time, we do not say that our spirit is prepared for the celestials only through qualities of things known to the senses, but also and much more through certain properties engrafted in things from the heavens and hidden from our senses. . . . It is for just this reason, that emerald, jacinth, sapphire, topaz, ruby, unicorn's horn, but especially the stone which the Arabs call bezoar, are endowed with occult properties of the Graces. And therefore, not only if they are taken internally, but even if they touch the flesh, and, warmed thereby, put forth their power, they introduce celestial force into the spirits by which the spirits preserve themselves from plague and poison.[47]

In his first major publication, *Propaedeumata Aphoristica* (1558), John Dee argued that besides familiar natural qualities the world is also filled with hidden energy, emitted from stones and metals, and also from words, songs, and so forth. This energy did not radiate to the surroundings through the known powers of perceptible forces such as light and fire, but rather through occult, animate traits, which were "living and perceiving," and could be employed, Dee believed, to the advantage of man.[48] Celestial harmony, Ficino had already explained, "oftentimes bestows a wonderful power not only on the works of farmers and on artificial things composed by doctors from herbs and spices, but even on images which are made out of metals and stones by astrologers."[49]

A Composite of Form and Matter

> For natural philosophers do not intend the image to be made of just any metal or stone, but of a certain one in which the celestial nature has initiated some time ago the power for what is desired and already almost perfected it.[50]

The act of carving an image on a gemstone or on a piece of metal, thereby injecting it with celestial influence to make a talisman, Ficino explained, was merely the culmination of a lengthy and gradual process in which the same influence was naturally conserved within the raw material. This ability to absorb and store celestial powers was shared by all terrestrial objects, yet different substances had different amounts of celestial energy invested in them, or released this energy at dissimilar rates. "[I]mages made of wood have little force," advised Ficino, "For wood is both perhaps too hard to take on celestial influence easily ... and it soon loses almost any vigor of cosmic life at all. ... But gems and metals, although they seem too hard for accepting a celestial influence, nevertheless retain it longer."[51] Metals and gems, he argued (in a way that somewhat resembles the use of the notion of "potential energy" in modern mechanical physics), could not be inserted into the belly of the earth without a considerable effort of the heavens, which effort was now embedded in them.

The role of the talismanic figure, on the other hand, was dispositive rather than directly causal, as it was designed only to dispose matter the better to receive and absorb celestial influence.[52] The beneficial or harmful astrological qualities did not reside in the image per se, but in the astrological conditions under which it was made. Nevertheless, just because a figure is immaterial it was allegedly more powerful in obtaining celestial virtues than physical objects. "Astrologers do not especially argue ... that our figures are the most powerful agents in themselves," observed Ficino, "but that they are the best prepared for catching the actions and forces of the celestial figures."[53] Terrestrial images, he argued, are subject to their celestial equivalents in heaven, provided that they are prepared at a time in which their correlating celestial forms (i.e., constellation or planet) are in their strongest influence. Sigils made of the appropriate metal, stamped at such a time with the symbol or image of the celestial body, caused the stellar rays to penetrate the material and create a new artificial object: "It [celestial nature] then finally perfects this power when this material is violently agitated by art under a similar celestial influence and begins to get warm from the agitation. And so art arouses inchoate power there, and when it has reduced it to a figure similar each to its own celestial figure, then forthwith it exposes it there to its own Idea; when the material is thus exposed, the heavens perfect it by that power with which they had also begun it."[54]

While our modern eye tends to look at images as mere representations, for Napier and his contemporaries they were also reproductions of the original (in this case the celestial object), partaking of its power. To understand the role of the figure in attracting celestial influence to the astrological talisman, Brian Copenhaver suggests we consider the Thomist concept of "substantial form."[55] In his *Summa Theologiae*, Thomas declared the heavenly bodies to be the principal active agents in the production of substantial forms. Unlike words or letters, he argued, which could be understood only as messages, images of celestial objects, when carved on the appropriate matter, became a natural feature of its being and an agent in the forming of its new substantial form.[56] Ficino built on Thomas to argue that the instantaneous act of carving an image on matter to make the astrological talisman created a composite object that was a member of a new species: that of the heavenly analogue of the image engraved. Thus, if a magus carved an image of a lion on a stone, the new artificial object became a member of the same species of objects that included the constellation Leo, and shared its substantial form.[57]

In his *On the Occult Works of Nature*, Thomas linked occult qualities with the imperceptible substantial form of matter, an association that had already been hinted at by Galen and developed by medieval Arabic writers. Ficino's obvious conclusion was that the new composite object was able to employ occult powers that were identical (if on a small scale) to those that emanated from the corresponding heavenly body. Celestial power was therefore received in an instant, but fused into the talisman in a way that guaranteed its radiation for a long time, allowing it to bestow the celestial gift on anyone who touched or wore it.[58] The implications for practical astrological medicine were enormous, opening new horizons for the natural treatment of early modern patients. Astrological talismans, correctly prepared and applied, became powerful remedies in the hands of able physicians such as Napier.

Making an Effective Cure

> If now any metal, in the time of its influence, . . . be stamp'd and hang'd about the neck, or carried, it will attract all antipathetick influences, and preserve men from approaching misfortunes, and clear them of all sicknesses.[59]

A Well-Timed Procedure

From the mid-1610s, Napier employed astrological talismans as part of his regular medical practice. The prescription of such a sigil was marked in his medical diary with a capital S followed by the astrological symbol of the planet the force of which it was designed to capture. The term "sigil," derived from the Latin word *sigillum* (seal), as well as its Ficinian synonym "image" (*imago*), referred initially to the figure carved or stamped on the talisman, but later came to denote in magical literature the composite object itself. It is not entirely clear why traces of Napier's employment of sigils start to appear only after Forman's death in 1611, but the proximity of the two events suggests some kind of linkage. Perhaps only after his mentor's manuscripts came into his full possession and so could be studied in depth, did Napier feel confident enough not to hide his engagement in such practices. Alternatively, this was simply a case of gradual vocational progression. Either way, the popularity of talismanic cures in Napier's medical practice peaked during the 1620s, only to decline considerably toward the end of his career. As I will show, astrological talismans became an integral part of his work as an astrologer-physician, an additional remedy enabling him to provide better care to his distressed patients.

"To make a sigil," proclaims Kassell, "was to stamp the powers of the stars into a piece of metal, creating an object both natural and artificial."[60] An effective astrological talisman was a powerful tool in the hands of the skilled magus; producing one required the careful following of exceedingly precise directions. Since, more than the timing of its employment, the conditions prevailing at the time of the sigil's preparation were crucial to its success in exploiting celestial force, a complex set of timing rules, extracted from careful a-priori calculations, guided the entire operation. To a lesser extent, practical directives also advised on how to choose and prepare the underlying metal or stone, in what substance it had to be stamped, soaked, or fumigated once prepared, what prayer should be said in the course of preparation, and how it had to be kept, carried about, or worn. All these factors had to be in absolute harmony, as an unknown magus explained: "He [the practitioner] ought to arrange all things according to the nature of their complexion, so that the mansion, metal, names of angels, characters, suffumigation, prayer, intention of his heart and the hour and day, and all things which pertain to the nature of his work come together in a harmony of quality."[61]

At the outset, the practitioner had to select the material of which the sigil was to be made. Both Napier and Forman occasionally used a gemstone, but much more commonly, most likely for practical reasons, prepared the sigil from a piece of metal. The choice of material depended upon the celestial force the influence of which was to be captured. In astral magic, each of the seven planets was traditionally twinned with one of seven metals, and a similar coupling existed for many of the known gemstones. Metallic-planetary associations, established sometime around the ninth century B.C. (based on former Chaldean influences), usually linked the Sun with gold, the Moon with silver, Mercury with quicksilver, Venus with copper, Mars with iron, Jupiter with tin, and Saturn with lead.[62] Nevertheless, other combinations existed, as one seventeenth-century writer admits: "We cannot know exactly under what *Planet* they [the metals] are subject wholly; . . . we are sure that . . . *Copper* [is subject] to *Jupiter* . . . [and] *Tin* to *Venus*, . . . Though, according to the opinion of several, *Tin* must be subject to *Jupiter*, and *Copper* to *Venus*."[63] Ficino advised his readers to prepare an image of Mercury on tin or silver, while in *Liber Lunæ* it was an image of Jupiter that was carved on a plate of the latter metal.[64] A sigil, like as any other astrological remedy, could also combine distinct celestial powers. For example, a sigil made from a mixture of copper and tin was prepared so that "in the on[e] syd[e] was ♀ and her caracts [characters] and in the other syde ♃ and his caracts [characters]."[65] Such a magical object could exploit the benefits of both celestial forces.

Elias Ashmole, the famous antiquarian who acquired Napier's diaries in 1676 from his grandnephew Thomas,[66] dedicated much time to the study of Napier's vast archive and, in particular, to some of his more occult practices. Over the next couple of years, Ashmole himself engaged in the casting of sigils and attempted to treat patients with them.[67] After a careful study of Napier's diaries, and perhaps to refine his own practice, he produced a very short but organized summary of the doctor's notes concerning sigils, meticulously referencing each instruction to its original location in the diaries. This two-page synopsis, containing fewer than a dozen items, identified each sigil either by its material or by the figure it bore. It included, among others, "The bigger (lamin) sigill of brass," "The little white sigill of tynn that hath the planet ♃ & his characters," and "a sigill of ♀ or ♃."[68] Considering the scarcity of detailed accounts or explanations about the use of astral magic in Napier's own records, and the fact that the cases are scattered among thousands of pages, this fairly short list provides much valuable insight into this practical art; I will return to it repeatedly throughout the following paragraphs.[69]

Napier sometimes prepared his sigils as constellated rings, to be worn on a patient's finger, but more often used rounded pieces of metal called laminas.[70] These were usually cast in order to serve as a direct therapeutic tool, applied by proximity to the ailing body. On other occasions, sigils were made as seals to be "sigillated" (i.e., impressed) onto a wax-like medicinal compound, which was then given to the patient to carry on his body or ingest. The composition of such medicine had to conform to the same celestial forces the influence of which was to be utilized, yet determining its specific ingredients and their exact ratio was less straightforward and did not rest on any clear or widely accepted recipes. "[F]or that the anncient wisemen, in the bookes that they haue written," complained one anonymous magus, "haue very much hid theire experince of the composition of the mass which must bee sigillated."[71] Describing the proper way to make a Jovian sigil, he was willing to share his version of this secret:

> [T]ake the powder of the most white frankincence, cloves, saffron, red corall, [and a] cristall stone called Cronioli, of each a like quantity in waight, and mixe it with oyle of roses or very good wine on the day & howre of ♃, and make therof a mass like soft wax, wherof make many round peecis of the quantity of the sigill of ♃, and let them be sigillated in either syde. then let them bee dryed in the shaddow, and putt them in a secret place vntill you haue occasion to use them.[72]

Napier's composition and use of such stamped medicines is revealed to some extent through Ashmole's summary, which outlined the doctor's recipe for sigillated medicinal "cakes": "Take of the pauder of [diagermis?] & diamargarit[on], aromat[ic] ros[e] ana ʒ [of each one drachm], sachar[um] ℥ [one ounce], rose water as much as will suffice, & sigillate with it. And give 2 litle cakes to serve her two mornings. give it in Brendi & not in any other liqueur, cold, at any tyme of the day."[73]

Astrological talismans characteristically bore symbols of the planets accompanied by seals of derived daemons and spirits,[74] symbols or images of the twelve zodiacal signs, or signs of other heavenly constellations. Also common were figures prepared according to the magic square associated with the relevant planet and its governing angel, or a diagram of the magic square itself.[75] Kabbalistic-astrological sigils usually bore the Hebrew names of the planet's archangel or intelligence, sometimes adjoined to Hebrew names of God. The accordance of the figures drawn to the celestial force that had to be exploited

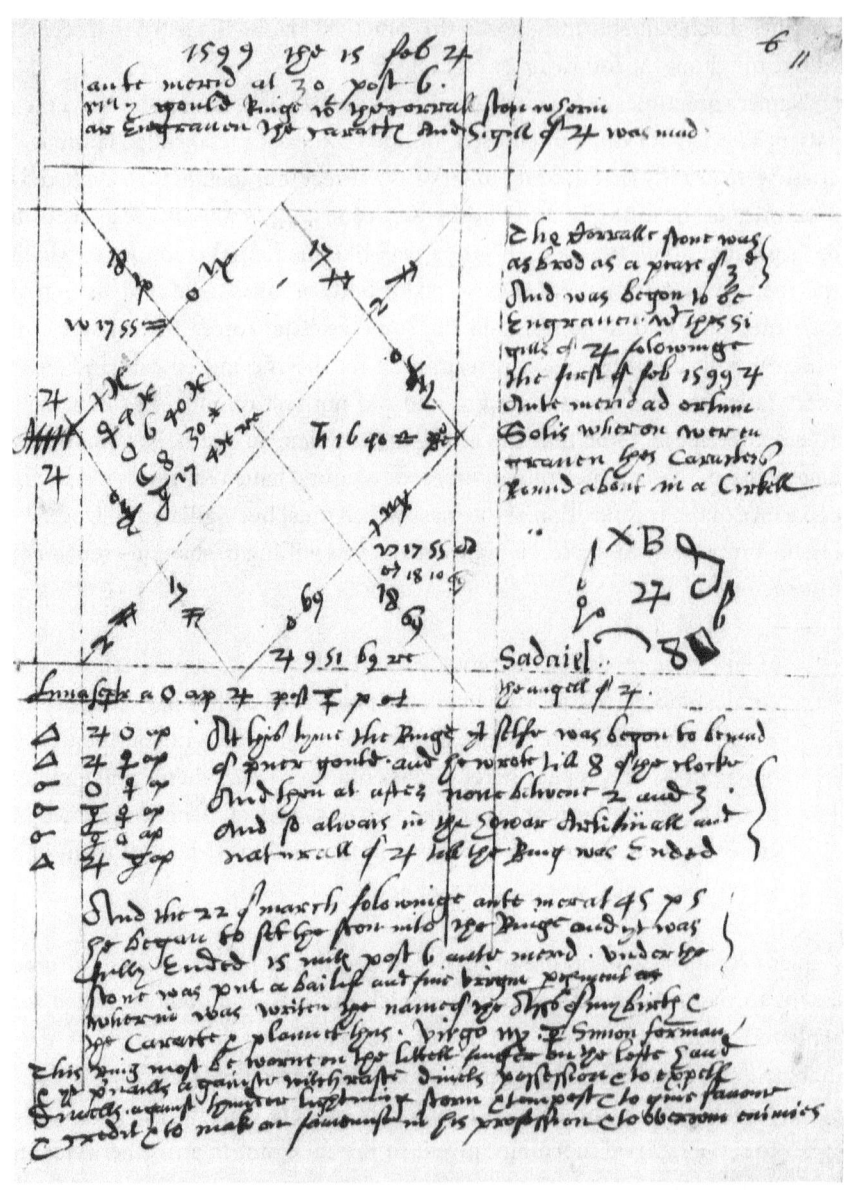

Fig. 3 Calculations for and description of the making and use of two gold rings with a coral stone, "wherin are Engrauen the caracts [characters] And Sigill of ♃," by Simon Forman. © The British Library Board, Sloane 3822, fol. 11.

was obviously the main concern in planning the sigil's face, but there were other considerations as well. "A sigill & characters on on[e] side," Ashmole quoted from Napier's notes, "is as effectuall as on both sides, & will serve on the one side very well."[76] Other decisions concerned the place designated for the symbols—in a circle or in the middle (in laminas), on the stone, underneath it, or on the ring itself (in a constellated ring)—and the proportion and size of the various signs and characters included.

Unlike Forman, whose manuscripts are scattered with dozens of astrological and kabbalistic designs for sigils, Napier's archive reveals nothing about the kinds of symbols or figures he used. Yet, from the traces of prescribed sigils and astrological correspondence and from the few references in Ashmole's précis, it seems that he usually restricted himself to symbols of the planets and perhaps the seals of derived spirits and daemons. This would accord with his attempt to refrain from any "demonic" or unnatural allusion that could have been suggested by the use of words, letters, or other intelligible characters.[77] Sigils were usually made by a craftsman (an "engraver"), and both Napier and Forman seem to have employed one.[78] A rather limited yet helpful set of their diary pages and instructions for the making and use of sigils, now bound in MS Sloane 3822, may cast some light upon how the actual manufacturing process of an effective sigil was conducted.[79] The entries record the type of sigil prepared, the precise time intervals in which it was cast or engraved, and an appropriate astrological chart for the time in which the entire process was initiated, often accompanied by a short astrological analysis. In a rather detailed entry, from the beginning of February 1599, Forman described the preparation of two golden rings set with coral stones and engraved with "the caracts [characters] and sigill of ♃" (see fig. 3). Next to the astrological figure, in which Jupiter is both *Lord of the Ascendant* and of the *planetary hour*, he noted:

> The corralle stone was as brod as a peace of 3^d And was begon to be Engrauen with this sigille of ♃ folowinge the first of feb[ruary] 1599 ♃ *Ante merid[iem] ad ortum Solis*. wheron wer[e] engrauen thes[e] caracters round about in a circkell.[80] Sadaiel the angell of ♃. At this tyme the Ringe yt selfe was begon to beread [be bred] of puer gould and he wrote till 8 of the clocke, And then at after none [noon] betwene 2 and 3. And so alwais in the howar Artificiall and naturall of ♃ till the Ring was Ended. And the 22 of march folowinge *ante mer[idiem]* at 45 p[ost] 5, he began to set the ston[e] into the Ringe, and yt was fully Ended 15 mnts [minutes] *post 6 ante merid[iem]*. [V]nder the stone was put a bailef [bay leaf] and some virgin parchment, wherin was written the name of the Acc[ident]s of my birth & the caracte[rs] & plannet thus, Virgo ♍ ♀ Simon forman.[81]

As I have explained earlier, the making of an astrological sigil culminated in the very act of engraving the figure upon the hard matter. Throughout this procedure, which typically took a couple of hours, the plain piece of metal or raw gemstone was believed to capture the celestial benefits presently dominating the skies. Performing this step within the proper time intervals was therefore crucial to the success of the whole operation, demanding the exact time frames to be carefully calculated according to the *planetary hours* (also called *artificial hours*) of the day.[82] Not least for this reason, the major part of Forman's careful documentation focused on aspects of timing. "Begin a little before 9," he advised in the making of a sigil designed to capture the benefits of Venus, "vntill 50 minuts after 9 in the morneinge, & in the after noone from a q[uarte]r after 4 to 10 minutes after 5."[83] Similar timing instructions recur in all of Forman's notes, sometimes adjoined by other considerations. In one example, the instructions for preparing a Jovian sigil list the usual astrological requirements for the success of the operation, but also add its desired whereabouts: "The sigill of ♃ is to bee graven . . . in the day and howr of ♃, ☽ beeing in with [sic] △ or ✶ aspect of ♀ or ♃, in a cleare day and without thy house, [and] without any curevture betwixt the skie and them, and a quiett ayre."[84]

Once the sigil was cast or engraved, it was sometimes fumigated (or "suffumigated," meaning fumigated from below) with the incense of specific herbs or animal parts in an attempt to purge it or to enhance its power.[85] On one occasion, after casting the sigils of Saturn and Mercury, Ashmole "took them out of the moulds and fumed them, with ♂ [antimony] and mastick,"[86] and elsewhere he even provided a list of "suffumigations" according to the planets.[87] Other sources suggest different versions. According to *The Picatrix*, for example, a Jovian talisman was to be suffumigated with a mixture of storax, peony root, aromatic calamus (sweet sedge), pine resin, the seeds of hellebore, and the feet of a dove. The *Speculum Astronomiae* recommended rather generally that sigils be fumigated with the wood of aloe, saffron, or balsam.[88]

Astrological talismans were made by casting the appropriate metal in pre-prepared molds and then manually engraving or stamping them with signs and forms relating to the celestial object the force of which was to be captured. As appears from the above examples, Napier's chief concern in this process was scheduling the act of carving to the astrologically propitious time he had carefully calculated beforehand. Only then could the raw piece of metal turn into a magical composite object, incessantly radiating the desired celestial influence. The complex instructions and limited "time windows" that guided their manufacture made astrological sigils valuable pieces of property.[89] Properly made, they became powerful and long-lasting tools of healing in the hands of the able astrologer-physician.

Sigil as Medicine

The seventeenth century witnessed a sharp growth in the medical applications of amulets and talismans. Israel Hiebner, an adherent in the latter years of the century, explained how sympathetic and antipathetic celestial forces served in the process of healing:

> I know also sufficiently, That all *Sicknesses* and *Diseases* in men, and other creatures, are occasioned by an *Antipathetick* Influence of the Heavens, Wherefore, when a Disease is known by the Nativity, you preserve your self, or totally cure it by an *Herb, Metal*, or *Stone*, gathered in due Season, and hang'd about your Neck, or carried about with you. Then, 1. The *Influence* of the Heavens and Times descend into the Creatures *Antipathetically*, or *Sympathetically*; that is to say, for a good or bad purpose. 2. When one of the above named three Creatures is prepared in due Season, and carried on the Breast, the *Antipathetick*, noxious Influence rather goes into the *Herb*, or *Metal*, than Men, and Man is preserved from the threatening Sicknes; but when the Sicknes is already in the Body, they extract it by degrees.[90]

Napier started to offer talismanic cures sometime in 1611, continuing their use throughout the rest of his career. During the mid-1620s, at the height of their popularity in his clinic, an astrological talisman was prescribed to nearly one in every ten patients.[91] Sigils were almost never prescribed by themselves, however, but rather accompanied a list of herbal remedies and other therapeutic treatments such as bloodletting or leeches (see fig. 4).[92] "If one be troubled in minde or tempted with ill thoughts, give a sigill of ♀ [copper]," suggests Ashmole, based on Napier's notes, "but purge first."[93] For patients haunted or bewitched, he quotes Napier's three-part instructions: bloodletting ("a little under the tongue, but very thriftily in the head veyne of the arme"), a purge with a diet drink (suited for the abounding humor), and a sigil.[94] Astrological sigils were therefore not only inherent to Napier's astrological-medical practice, but also naturally integrated within his eclectic collection of therapeutic offerings. Given primarily as one component within a whole treatment, they comprised an additional curative tool within his medical practice.

When applying a talismanic cure, the first and foremost decision for Napier was matching the appropriate sigil to the patient's specific illness. Most astral sigils were related to a single governing planet or constellation, encompassing the heavenly power of that celestial object. "When this *Sigil* is prepared in the influence of *Jupiter*, and put in a blue Silk bag," advised Hiebner on the use of a

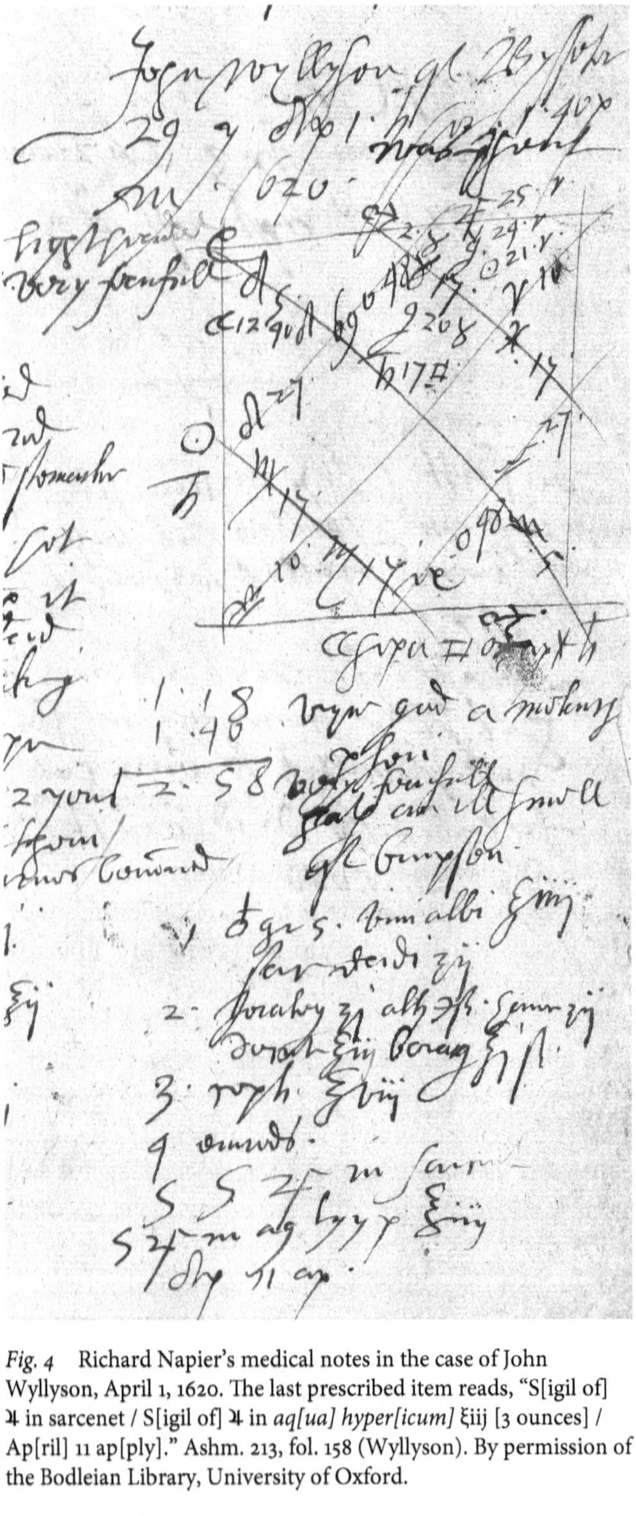

Fig. 4 Richard Napier's medical notes in the case of John Wyllyson, April 1, 1620. The last prescribed item reads, "S[igil of] ♃ in sarcenet / S[igil of] ♃ in *aq[ua] hyper[icum]* ʒiij [3 ounces] / Ap[ril] 11 ap[ply]." Ashm. 213, fol. 158 (Wyllyson). By permission of the Bodleian Library, University of Oxford.

copper or tin talisman, "it cures all the *Sicknesses* of *Jupiter.*"⁹⁵ A Jovian sigil, therefore, was essentially the talismanic equivalent of a Jovian herb, concoction, or mineral drug, and was designed to cure diseases that were traditionally treated through the sympathetic or antipathetic influence of Jupiter.

Some practical directives linked sigils, made at certain astrological states, to ailments, defined by more specific combinations of "Planet in sign," according to the same pattern I presented in chapter 1 for Napier's mainstream astrological physic. "[The] Sig[ils of] ♃ in ♎ in 10, ♌ 7 in ♎ in 10," reads one of Forman's rules, "are good against all diseases of ♃ & of ♄ in ♒." "[The] S[igils of] ♃ & ☿," advises another, "haue vertue & power to cure the diseases caused of ♄ [in] ♓."⁹⁶ A note made by Ashmole in his diary of October 1672 sheds some light on Napier's considerations in attaching a sigil to a specific medical condition: "Sir Richard Napier told me, that the ring which his uncle made for the falling sickness, was after this manner: He took the ascend[an]t when the patient came to him, that was soe afflicted, & vnder the same ascend[an]t made a silver ring."⁹⁷

Choosing the appropriate sigil was Napier's foremost concern, yet there were other practical considerations to be pursued. These concerned the time of application (when to wear the sigil, sigillate with it, or consume the sigillated medicine),⁹⁸ the manner of wearing (e.g., wrapped in silk, or if a ring, on what finger),⁹⁹ and the prayer to be said (before wearing or sigillating).¹⁰⁰ Magical practices often resembled or borrowed from religious rituals, and acts of purification were sometimes required not only of the object itself but also for the practitioner or patient: "Because such *Sigils* are stamp'd at such times when the Planets are in their highest Dignities, therefore it is here to be noted, that he who uses them, must not use them when he goes into the *Bath*, or lies with a Woman, or [goes] into unclean Places."¹⁰¹

Eager to learn the rules of talismanic healing, Ashmole listed, in his aforementioned summary, the infirmities to which each of Napier's sigils applied, occasionally noting additional guidelines for its administration: "The little white sigill of tynn that hath the planet ♃ & his characters, is good for many infirmities, for all manner of disease, for falling sicknes, gout, dropsis [dropsy], apophisees [apophysis], stone, chollick ["wind colic,"] or any other disease, whatsoever, curable or incurable, with good prayer. And also against all evil spirits, fairies, witcheries, possessed, frantick [and] lunatick."¹⁰² This may appear a surprisingly general directive, but it seems to correctly reflect Napier's practical considerations. In fact, both Ashmole's retrospective notes and the relatively frequent use of a Jovian sigil in Napier's consulting room (as compared to the frequency of administering any other specific medicine) suggest that certain sigils were used in a broader way than most of their herbal equivalents.

Except at the very beginning of his astromagic career, Napier equipped his patients almost exclusively with the sigil of Jupiter. Its prescription, marked with an "S ♃," is frequently followed by one of four combinations: "in *aqua hypperi-cum*,"[103] "in taffeta,"[104] "in sarcenet,"[105] or "appensile."[106] The nature and function of these terms are not always clear. *Hypericum perforatum* (the Latin name of the medicinal herb St. John's wort, derived from a Greek word meaning "over an apparition") was used in Christian folklore as "a talisman against evil."[107] "In *aqua hyppericum*" probably meant steeping the sigil in the distilled waters of the herb.[108] This act of soaking the sigil or the sigillated tablet in liquid medicine served as a purge or to enhance power. Very often, the charmed potion itself was given to the patient to swallow. For instance, Ashmole learned from Napier that a trochee, sigillated with the "little white sigill of tynn," was to be "steeped in drinke, or wine or milk, or in loringes" before its consumption.[109] Taffeta and sarcenet were two types of silk in which the sigil was probably wrapped. This is consistent with other evidence of talismans wrapped in silk or in other fabrics.[110] "Appensile" I found quite indecipherable, but, with some effort, it might be identified as the noun form of the archaic verb "to appense," meaning "to be hung from above" or "to append a seal." Perhaps an "appensile" was a small satchel hung from the neck, or a term applied to a sigillated medicine.[111] Further directions were sometimes supplied for the ribbon on which the sigil was to be hung. In one of these instances, Napier prescribed for a patient a Jovian sigil "made in taffat[a] with a fayne silke ryband."[112]

In chapter 1, I showed how Napier measured the success or failure of his medical treatments by evaluating the outcome of induced bodily excretions. Since astrological sigils had no equivalent observable outcome, such an assessment could hardly be carried out.[113] Nevertheless, Napier was always happy to record a client's positive feedback: "Henry Mun of Wolverton came to signify that his sone that hath bene so lame & could not styrre [stir] for a long tyme is by the S[igil] of ♃ so recovered, that by gods *gratus* favour he now goeth abroad. he came to give me thankes."[114] Such appreciations were not restricted, of course, to the outcome of talismanic remedies, but they certainly highlight the practical value of these enchanted objects for Napier.

Natural Efficacy, Superstition, and the Lawfulness of Images

A Lawful Art?

"Things that have no cause either in nature or by divine institution," wrote Napier's contemporary the Lutheran theologian Ludwig Dunte, "are performed by magic."[115] Thus defined, the tagging of his own work as "magic" would have been utterly rejected by Napier as both erroneous and misleading. Aware of the possible implications of cunning, and intellectually equipped to recognize the wider context of his practical techniques, Napier considered himself a practitioner of natural philosophy and an explorer of nature. In exploiting wondrous celestial influences, he believed, he merely utilized the great forces of the cosmos via perfectly natural means. Throughout the seventeenth century, however, "Puritan" theologians wrote zealously against the employment of magical remedies and attacked cunning men and women as mere witches. "These vnlawfull and absurde meanes are more vsed & sought for of common people, then good physicke," complained the prominent pro-Calvinist theologian William Perkins; "It were better for a man to die of his sicknes, then to seeke recouerie by such wicked persons."[116] More "conservative" churchmen, however, were less concerned. The future archbishop of Canterbury, William Laud, showed only slight interest in the subject of magic, and "told the Duke of Buckingham around 1625 that magical healing was a topic he had 'little looked into.'"[117] Even pious patients did not necessarily link religious and medical conformity, and some resorted in their distress to the rather controversial methods of popular magi. Indeed, some "Puritan" clergy did not hesitate to seek Napier's medical aid. John Williams, dean of Westminster and later lord chancellor and archbishop of York, came to see him in April 1620 with a cold and "stuffed about the head," and William Twisse, the vicar of Newbury, who publicly attacked Napier in 1617, had come to consult him a couple of years earlier about his sick wife.[118]

The Christian debate over the lawfulness of magical healing dates back to the earliest church fathers. What would become the orthodox hard line was given by Augustine in his *De Doctrina Christiana*: "It is one thing to say that if you drink this herb after it has been crushed, your stomach will not hurt, but quite another to say that your stomach will not hurt if you hang the herb about your neck. The one is approved as a preparation that brings health; the other is a token of superstition and condemned."[119] The medieval distinction between natural and demonic magic helped alleviate this orthodox position and made way for the possible authorization of some magical techniques. The *Speculum Astronomiae* identified three different kinds of talismanic magic. The first, which was astral in nature but employed invocations or other ritualistic actions, was tagged "abominable"; the second, employing characters, names, and exorcisms, was tagged "detestable"; whereas the third, which relied only on stellar powers, was considered lawful.[120] This functional classification, maintains Nicolas Weill-Parot, later guided the Thomist distinction between explicit and implicit pacts with demons, but also allowed certain kinds of employment of astrological images to be assigned to the third category of licit magic.[121]

The seventeenth-century debate over the legitimacy of magical practices therefore revolved less around magic per se than the issue of what was or was not "natural." The Christian church itself never objected to attempts to heal the sick by purely natural means, but any claim to have achieved something that was not based on acknowledged natural causes was immediately suspect.[122] For Napier and his contemporaries, however, the dispute over what was natural was hardly an easy one, for it touched upon problematic religious issues. Building on the works of Jean Delumeau, Robert Muchembled, and Keith Thomas, Stuart Clark argues that Protestant reformers undertook a program for the "Christianization" of the laity. Their aim was to destroy what they believed to be the common people's "superstitions," which were defined as "false practices attributed to false causes" or the association of causes and effects that did not belong together in nature. What was faulty in nature, argues Clark, was consequently faulty in religion and therefore in need of correction.[123] Cures such as those offered by Napier, therefore, were not deemed superstitious because they did not work, but rather because of the manner in which they supposedly *did* work, namely, through an inappropriate appeal to supernatural forces (i.e., the devil).[124]

The difficulty in differentiating between an efficacious natural action and a suspected superstition lay in the many obscure virtues in nature, which could produce apparently causeless effects. Anyone who was prepared to follow Ficino in accepting a causal relationship between celestial influence, astrological sigils, and human health, as was Napier, could defy the tagging of astral magic as "superstitious" and recognize the natural (and hence lawful) framework of its practical acts.

The Problem of Graven Images

The dispute over the lawfulness of images was rooted in the intense Protestant campaign against the employment of religious paraphernalia as tools of worship. In the first half of the sixteenth century, Reformers such as Martin Bucer and Peter Martyr on the Continent, as well as Thomas Beccon and Thomas Cranmer in England, advanced a new reading of the Decalogue in which the prohibition against images was made a separate second commandment.[125] Exploiting the ambiguity of the original biblical approach to the question of using images, these zealous theologians stepped "from the wrongfulness of worship to the wrongfulness of making and having images," focusing particular attention on three-dimensional decorated objects.[126] John Calvin, their theological mentor, asserted that the root of idolatry lay, not in the pretension to create a man-made deity, but in the attempt to invoke divine benefits through the visual representation of heavenly forces. It was this conviction of the presence of power in the image or decorated object that stood behind their removal from Reformed churches.

Napier's original apology for the lawfulness of using graven images could therefore be read as a purely theoretical theological essay.[127] Its reference to the images in Solomon's temple, the ultimate prototype of the Christian prayer house, seems to reinforce such a notion. Yet in view of the "Puritan" campaign against images, some of Napier's enemies were only too eager to stress the dangers of idolatry inherent in his astral talismans.[128] Although the text appears to deal with statues and monuments rather than astral images or talismans, Napier's general reluctance to engage in perilous theological disputes devoid of any practical implications, together with the tract's particular stress on three-dimensional objects, suggests that it might be more appropriately read as a defense of the practical use of carved images.

The tract opens with a historical and biblical survey, intended to establish the foundations of the use of images and sculptures in the Judeo-Christian heritage. "The antiquity & originall[ity] of images," asserts Napier, "some ascribe to Enoch, others to Prometheus, others to Ninus, others to Nahore.... Epiph[anius] thinketh that he [Nahor] was the first that made the image of his sone, of Haran."[129] Ascribing the making of images to such ancient authorities might indeed demonstrate the antiquity of the practice of using images, but not its lawfulness, as not every biblical reference was necessarily made in a positive context. In fact, admits Napier, the images of Haran were idolatrous, prepared "through pride & ambition, tyrany, blind ignoranc, fonde superstition & the subtilly of Sathan worshipped as god."[130] Nevertheless, he insists, a distinction between lawful and unlawful images exists, and its nature is rather straightforward: "it is not lawfull to make an idolatricall image, yet it is lawfull to make other images which be not idolatricall."[131] Napier's division employs the practical aim of the operation in order to test its lawfulness: a position that rightfully emphasizes the intent of worship, but which would probably have been rejected by his godly opponents as absurdly naive. Aware perhaps of this problem, Napier repeats the same argument, this time making a direct reference to the disputed second commandment. "The making of images is not by this command[ment] simply condemned without all respect," he explains, "but with respect as it is made to be an idoll, or to be worshipped as god or in steede [instead] of god, either voyding or robbing god of his honour."[132] To stress this point further, Napier focuses on biblical examples in which images were clearly used idolatrously.[133] When it comes to those "that serve graven images that glory in idols," he concludes decisively, "boath the image & he that made it is as cursed," and the images should be shattered to pieces or burnt.[134] "[T]he bare & simple making of images," on the other hand, "are [sic] not prohibited by this commandment."[135]

Expounding the genuine intention of the second commandment, Napier challenges its misreading as a ban against the use of any kind of image. As a matter of fact, he argues, a number of biblical tales positively attest to the lawfulness of images and ornaments. The book of Kings, for example, teaches us that "Salomon sent for Hiram to worke all maner of work of brasse, . . . & he is not condemned for it."[136] The same Salomon "builded a temple to god which was decked with all maner of pictures & had the images of lions."[137] Clearly, such an allusion could not only legitimate the employment of astrological sigils, but also justify tolerance for the use of images inside churches. It is in this context that Napier deals with another distinctive kind of "graven image," that of the historical monument. "As for historicall & politicall images it is certen they be not prohibited,"[138] he argues, based on a famous passage from Matthew and several other scriptural accounts of the construction of monuments.[139]

Napier's arguments by example as given hitherto were too weak, he felt, to establish a strong stand against religious extremism. He therefore moves on to present, toward the end of the pamphlet, a couple of more essential claims. Reflecting the Renaissance notion that all natural things, including man-made arts, are inspired by God, Napier asserts, "[I]f god give an art, then the vse of that art must needes be lawfull."[140] Such line of argument allows him to launch a general defense of the lawfulness of exploring nature. The second claim is developed by way of an epistemological argument concerning the essence of God: Napier presents two biblical quotes, from Isaiah and 1 Timothy, to show that God is invisible and cannot be likened to anything: an immortal being which "never man sawe, neither can see."[141] Since God cannot even be imagined, much less reflected in an artificial object or figure, the idea that he can be represented by one is simply absurd: "No man hath seene God at any time. . . . Therfore he cannot be represented by any image or similitie whatsoever. he is infinite & incircumscriptible, immortal [and] everlasting, & therfore he cannot have any leeknes [likeness] or image, all which be finite, mortall & fading. Images are the remembrances either of things dead or absent but soe it is not with God."[142]

This extraordinary apology by Napier provides a rare opportunity to learn, at least to some extent, how he reconciled his generally orthodox religious outlook with some of the more obscure aspects of his practical medicine. At the same time, it sheds some light on his fears and concerns as a practitioner of disputed arts, confronted by "populist" criticism that time and again threatened to disrupt the busy serenity of his medical clinic. Indeed, what might have been seen by others as potentially unlawful (or at least suspect) magic embodied for Napier a powerful technology of practical medicine that fitted fairly neatly within a Christian-Neoplatonic view of the cosmos. By his own light he remained both the magus and the orthodox Protestant cleric.

3

CONVERSE WITH ANGELS

דא רפאל, דאיהו ממנא לאסוותא דארעא, דבגיניה אתרפיאת ארעא, וקיים בר נש עלה, ורפי לכל חיליה

This is Raphael (lit. healer of God), who is charged to heal the earth, and through whom the earth is healed so as to furnish an abode for man, whom also he heals of his maladies.
—*The Zohar*, 1:146 [46b–47a] (original in Aramaic)

History and Reputation

On a trip to Oxford in the summer of 1667, the biographer and antiquary John Aubrey (1626–1697), a member of the Royal Society, became acquainted with Anthony à Wood (1632–1695), a local historian and fellow antiquary, with whose brother Aubrey had studied at Trinity College. Wood was seeking to write a biographical dictionary of Oxford writers and ecclesiastics, but his alienating character and his remoteness from London had brought him certain difficulties, with which Aubrey offered to help.[1] Aubrey's pen was notorious for its freedom, and his notes included the latest gossip, but also valuable information collected from unpublished manuscripts. One of his favorite resources was Elias Ashmole's vast private collection, which incorporated material from various personal archives, including those of John Dee, William Lilly, John Booker, and Richard Napier. Rather than an isolated, private space, Ashmole's library was an accessible repository of information, attracting scholars from all around the country, and a true treasure trove for a seeker of life histories such as Aubrey.[2]

Ashmole himself, fascinated with Napier's occult practices and extraordinary skills, had put much effort into organizing, indexing, and summarizing many of his manuscripts.[3] Yet in the middle of the year 1681 Aubrey noted with disappointment that his older colleague seemed to have withdrawn from an earlier promise to write "the lives" of Napier and several other "remarkable persons."[4] It was up to Aubrey, then, to provide the impetus to ensure that the late doctor received the publicity and recognition he deserved. From December that year, this diligent antiquary made frequent visits to Ashmole's study, joining him in making "a collection of Nativities of learned men," based largely on the manuscripts of Napier and Lilly. Aubrey regularly updated Wood on his efforts at the archives. "I will after Xmas goe most Days to him," he wrote in one of many letters, adding that the two men hoped by April "to turne over 40 volumes." "Mr Ashmole turnes and reads," he said, describing the joint endeavor, "& I doe write."[5]

Seven years later we find Aubrey discussing with Ashmole the transfer of his own collection of manuscripts, some of which were then in the possession of Wood, to the newly founded Ashmolean Museum in Oxford. Aubrey sent Wood a series of letters urging him to hand over the papers to Robert Plot, the first curator of the museum, and complaining of his continued refusal to do so. In one of these pleas, perhaps to slightly temper the otherwise harsh tone, he also updated his colleague on some of his noteworthy findings. "About the middle of the Dreams," he wrote, "you will find a very remarquable passage concerning Doctor Napiers foretelling Doctor Prideaux his being a Bishop . . . , as also concerning Sir Geo[rge] Booth (first Lord Delamer)."[6] These two curious hints of Napier's conversations with angels featured in the brief accounts of his life that later appeared, almost identically, in both Wood's *Athenae Oxonienses* (1692–93) and Aubrey's own *Miscellanies* (1696).[7] As I will show, Aubrey's complete account included much more than these two appetizers. Rather, it was a well-founded narrative that would be much repeated by later scholars and so become the core of Napier's long-lasting reputation as a conjurer of angels.

A second biography of Napier emerged from a slightly longer yet fragmented report in William Lilly's celebrated autobiography. Written in 1668 at Ashmole's request (but not published until 1715), Lilly's work is famous for its remarkable depiction of the era's community of astrological practitioners. Less noticed, at least until quite recently, is its similar portrayal of contemporary conjurers and seers.[8] Many of Lilly's most vivid reports describe astrologers, diviners, and healers who were involved in conferences with angels and spirits, hinting at the extent and diversity of this occult practice. Lilly mentions, among others, his own mentor in astrology, John Evans (and his daughter Ellen), the astrologer William Hodges, an apothecary by the name of Charles Sledd, Sarah Skelhorn (a speculatrix to the physician Arthur Gauntlet), and the famous John Dee.[9] These men and women summoned angels and spirits and requested their celestial assistance in various worldly matters, in a context that Joad Raymond interprets as "an aspect of astrological practice."[10] Lilly's account of Napier, on the other hand, portrays him as an able astrologer-physician and devout churchman, but makes no mention of his angelic contacts, with but one exception—upon the writer's visit to Great Linford, Lilly records, Napier "invocated several angels in his Prayer, *Viz. Michael, Gabriel, Raphael, Uriel, &c.*" Ashmole, who edited and annotated Lilly's autobiography, remarked on this statement: "*At sometimes, upon great occasions, he had Conference with Michael, but very rarely.*"[11] The general silence regarding Napier's angelic conversations was only surmounted by Aubrey's sustained work on the manuscripts.

Aubrey and Lilly provided two distinct historiographical narratives of Napier that subsequently traveled through the ensuing centuries, sometimes in parallel, sometimes intersecting. In the absence of any further serious examination of his life and work until late in the twentieth century, they were jointly responsible for Napier's long-lasting name as a devout rector, skillful physician, and, most significantly, seer. In what might resemble Aubrey's efforts with the archives, this chapter will examine in depth the practical facet of the doctor's intriguing conversations with angels; but first I would like to trace the evolution of Napier's reputation, which thrived on the continual clash between those who accepted the reality of such spiritual communication and those who sought to condemn it as mere superstition.[12] It is worth noting in this context that categories of superstition and reason are complex, and while the suggestion that humans could engage with spirits was already suspect when Napier made his communicative efforts—as was also, albeit to a lesser extent, the possibility of knowing anything about them—an almost undisputed consensus surrounded the belief in the actual existence of various heavenly creatures.[13] Still, as late as the last quarter of the seventeenth century, it seems that Napier's younger admirers were ready to accept as trustworthy not only his prophetic insights but also their alleged source.

Aubrey's *Miscellanies*, as its title suggests, is a seemingly random collection of various accounts. Its numerous anecdotes of supernatural phenomena, drawn from the writings of church fathers, travel writers, old manuscripts, and "tradition," are divided thematically into sections entitled "Dreams," "Apparitions," "Omens," and so forth. Under the title of "Converse with Angels and Spirits," Aubrey sketched the story of the once-renowned doctor: "Dr *Richard Nepier* was a Person of great Abstinence, Innocence, and Piety: . . . When a Patient, or Querent came to him, he presently went to his Closet to Pray : and told to admiration the recovery, or Death of the Patient. It appears by his Papers, that he did converse with the Angel *Raphael*, who gave him the Responses."[14] Aubrey's account rests, as I have just shown, on a close examination of Napier's papers, which were then in the possession of Ashmole. However, nothing in the manuscripts themselves appears to testify explicitly that the doctor actually communicated with angels. Attitudes toward the summoning of angels in the early seventeenth century were rather ambivalent, but explicit law forbade the conjuration of spirits.[15] This was probably enough to deter Napier, never very keen on publicizing his occult activities, from leaving behind any "incriminating evidence." Considering this lack of direct textual proof, Aubrey's allegedly anecdotal tale is revealed at second glance as a reasoned, even convincing story. Woven throughout his text are references to four different kinds of evidence he had considered: a text of invocation ("a call"), the ℞ sign, some examples of personal and general queries, and the direct testimony of one of Ashmole's acquaintances.[16]

The most comprehensible traces of Napier's angelic contacts survive in several dozen folios, assembled toward the end of most of his medical casebooks, from the early 1610s to the mid-1620s. These papers, which I refer to here as the *Interviews*, contain hundreds of short or elaborate queries and their responses, sometimes prefixed with a curious ℞ sign (see fig. 5). With answers jotted hastily besides some of the questions, often in a different ink, they appear as undigested notes set down in the course of practical divination.[17] Aubrey's first anecdote, concerning the episcopal aspirations of the ecclesiastic John Prideaux, was picked up from one of these pages.[18] A note made apparently in the summer of 1619, records Napier's inquiry "Wheather D[r] P[rideaux] will live to obtayne his pretens as to be a deane & a bishop of Oakley."[19] The answer, written in a different ink, assured that "he will." About three months later, on a different sheet, the same prophecy is repeated: "D[r] Prideox will live to be a bishop."[20] As I will show, most personal questions in the *Interviews* concerned either specific medical cases or major life concerns such as life expectancy, marriage, children, and heirs. The assessment of Dr. Prideaux's personal aspirations was therefore quite extraordinary, but it did serve Aubrey as a good illustration of Napier's exceptional skills.[21]

Aubrey's second example concerned the prophesied birth of Sir George Booth (1622–1684), an active parliamentarian in the Civil War and a member of several English Parliaments. It was his future father, William, who came to see Napier, and "R ℞is, did resolve him," reports Aubrey, "that Mr. *Booth* ... should have a Son that should inherit, Three Years hence ... *viz.* from 1619."[22] Napier's *Interviews* contain indeed such a prophecy from September that year: "Mr Booth shall live to [be] 40 y[ears] & have within these 3 y[ears] a boy who shall inherit."[23] Aubrey was therefore faithful to his sources but, as I will show, selected only the curious and remarkable titbits from what seems to have been a much broader, sustained effort by Napier to gain celestial wisdom.

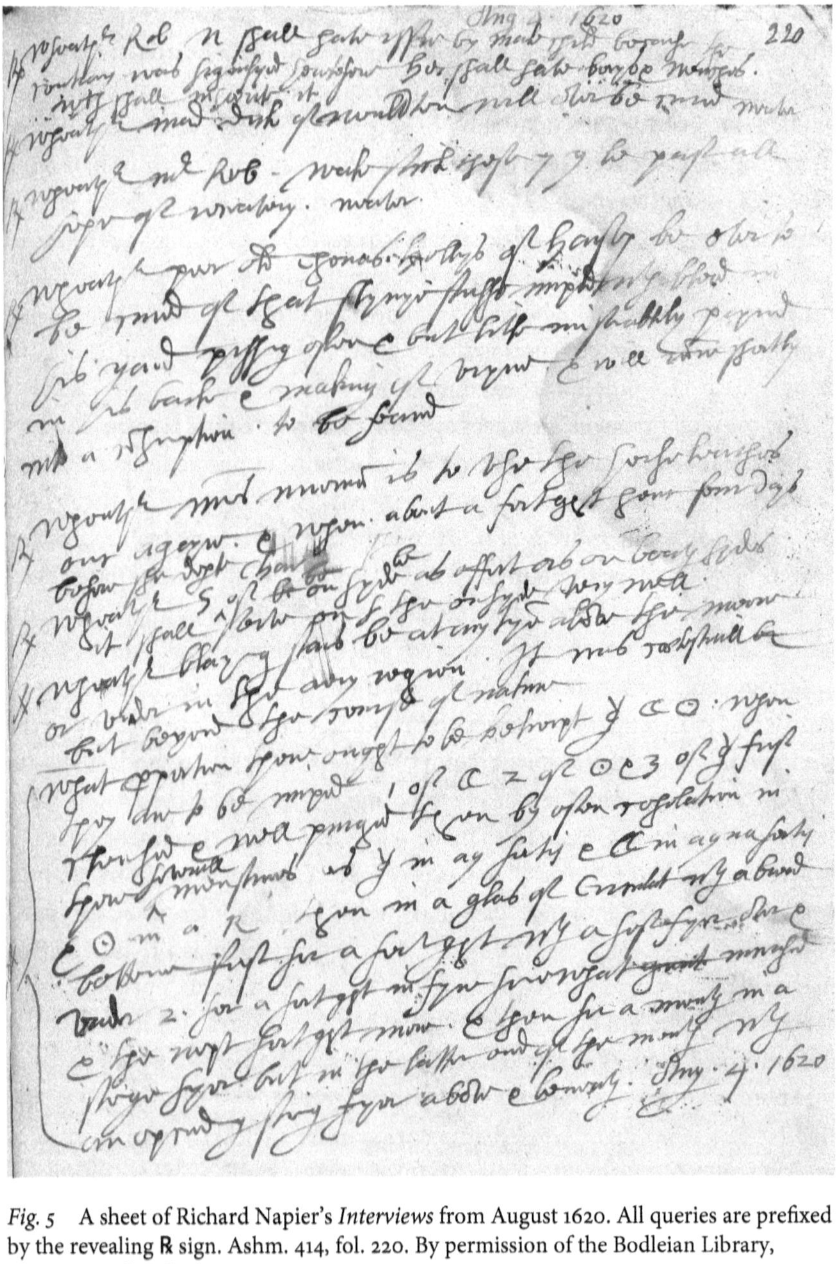

Fig. 5 A sheet of Richard Napier's *Interviews* from August 1620. All queries are prefixed by the revealing R sign. Ashm. 414, fol. 220. By permission of the Bodleian Library, University of Oxford.

The key to Aubrey's report was the intriguing ℞ sign, which, as he rightly remarked, stood before many of Napier's queries, prognostic notes, and medical prescriptions, and "which Mr. *Ashmole* said, was *Responsum Raphaelis*."[24] This symbol, which we identify today as the unmistakable mark of a medical prescription, has a long history in medicine, preceding Napier by many centuries. In ancient Egypt, the sacred eye of Horus, or "the eye of Ra," represented by the letter *R* and an eye enclosed by a circle, was known as a symbol of protection and healing and a token of royal power.[25] Around the first century A.D., Roman physicians began to use the sign of Jupiter in prescriptions to indicate that they were operating under the authority of Caesar and the protection of the gods. Over time, the signs of Horus and Jupiter have combined to form the sign of ℞ (or Rp), which continued to denote medical recipes. Napier, however, avoided the use of this graphical mark in prescriptions, and it is missing altogether from his notes before and after his engagements with angels.[26] Rather, it visually distinguishes his communications with Raphael from the rest of his documented medicine and from that of his contemporaries.[27] Scattered throughout his diaries, mostly during the early 1610s, the ℞ sign, together with the initials of other angels, seems therefore to validate the celestial origin of his numerous prophetic, therapeutic, and diagnostic remarks.[28]

Aubrey finds in the testimony of a gentleman named Edward Waller, an acquaintance of Ashmole, further confirmation of Napier's alleged conferences: "When E. W. Esq; was about Eight Years old, he was troubled with the Worms. His Grand-father carried him to Dr. *Nepier at Linford*. Mr. E[dward] W[aller] peeped in at the Closet at the end of the Gallery, and saw him uppon his Knees at Prayer. The Doctor told Sir *Francis* that at Fourteen Years old his Grandson would be freed from the Distemper; and he was so. . . . 'Twas about 1625."[29] Aware, nevertheless, of the apparent inconclusiveness of his circumstantial evidence, Aubrey wards off potential criticism by distancing Napier's practice from alternative modes of divination. "It is impossible," he asserts, "that the Prediction of Sir *George Booth*'s Birth, could be found any other way, but by Angelical Revelation."[30] In fact, both the prognostic remarks in Napier's medical diaries and the prophecies voiced by the *Interviews* could have been concluded from other popular methods of early modern divination, specifically astrology. However, while Aubrey's proofs might fail to meet today's scientific standards, they did establish a legitimate reading of authentic sources for his readers.[31]

The *Interviews*, judging from those that ended up with Ashmole, reflect Napier's use of angelic communication as a practical tool in both medical practice and divination. Holding arrays of questions of varied lengths, they touch upon the cause, cure, and outcome in specific medical cases, personal anxieties and hopes, and general knowledge of medicine, religion, and alchemy. Some sheets concentrate on a specific theme, in what appears to be pre-prepared lists of questions coupled with the archangel's later-recorded replies; others combine queries from different domains or contain the dictation of rather lengthy recipes. Unlike Napier's medical diaries, which provide continuous documentation for his entire medical career, only a few dozen folios from the *Interviews* survive, covering a handful of different dates.[32] It is reasonable, nonetheless, to assume that they are the random remains of a much wider documentation, lost or deliberately destroyed by Napier himself. Aubrey's report provides in this case a glimpse into what was a much larger corpus of angelic conversations and a frequent, regular technique embedded in Napier's practical medicine.

The first to quote Aubrey was the eighteenth-century bookseller and autobiographer John Dunton, the author of numerous political pamphlets supporting the English Whigs. In one of these booklets, published in 1714 and titled *Dunton's Ghost*, Dunton builds on Aubrey's account of Napier in arguing that humans can receive information from angels and spirits. "'Tis plain by this Instance of Dr. *Nepier*," he proclaims, "that some good Men have received News *from the World of Spirits*, by setting a Correspondence with ANGELS and GHOSTS. . . . [F]or certainly none can doubt, but those intermediate Agents (the *Angels*) are employed Reporters, and Transporters, Monitors, Couriers, [and] Guardians, between this and the other World."[33] Another early reference to Napier's angelic conversations appeared in Samuel Madden's intriguing *Memoirs of the Twentieth Century* of 1733. This opus of six volumes provided a prophecy of the "most Important Events in *Great-Britain and Europe*, . . . From the Middle of the Eighteenth, to the End of the Twentieth Century, and the World," compounded as a series of "letters from the future" its author had supposedly received from his guardian angel. Napier and Dee close a second preface—a weary defense of the existence of guardian angels and their ability to communicate with men—as a couple of noteworthy English examples.[34] These and other accounts of Napier in the 150 years after his death were generally favorable, building mainly on Aubrey's report to confirm the possibility of angelic communication.

An early harbinger of a change of tone and the tagging of Napier as a "superstitious fool" was Daniel Lysons's 1792 *The Environs of London*, which later developed into the *Magna Britannia*—a comprehensive survey of the history of some of the counties of England written jointly with his brother Samuel. Lysons extracted most of his facts on Napier from Lilly's *Life*, but also claimed that he "conversed with the angel Raphael."[35] His report already expressed the criticism that we will encounter in later accounts of Napier's angelomancy. "We have had empirics and enthusiasts of late who have professed to cure diseases by means as extraordinary, and who have had their pretended conferences with angels," complains Lysons, "nor have there been wanting those who have been credulous enough to listen to them." Directly referencing Napier, the brothers' *Magna Britannia* was somewhat more subtle, yet still looked upon the doctor's spiritual practice with a cynical eye: "His practice as a physician became very extensive, it being given out that he held conversations with the angel Raphael. ... This procured him great credit in a superstitious age."[36]

A review of the *Magna Britannia*, published in the 1808 issue of the *European Magazine and London Review*, combined the brothers' criticism with an outline of Napier's life and work, based on bits of information extracted from both Aubrey and Lilly.[37] The new, critical attitude became even more manifest in a short article published about a year later in the spring issue of the *London Magazine*. Part of a series of biographical pieces titled "Additions to Lord Orford's Royal and Noble authors," this particular item discussed the life of Sir Henry Booth, citing Aubrey's story of his son's prophesied birth. The author, Philip Bliss, who edited Aubrey and Wood and published some extracts from Forman's manuscripts, discussed what he termed "Aubrey's superstition and credulity," and offered his fantastic account of Napier as yet another example of misguided writing.[38]

The first half of the nineteenth century, with its increased interest in esoteric folklore and ancient superstitions, abounded with histories of demonology and magic.[39] One of the most famous works in this area was Sir Walter Scott's *Letters of Daemonology and Witchcraft* of 1830, which surveyed, in the form of letters to Scott's son-in-law, the extensive history of magic and witchcraft. Taking a rigorously rational approach to the subject, Scott attributed supernatural visions to "excited passion," credulity, or physical illness.[40] Another celebrated work was George Conrad Horst's six volumes of *Zauber Bibliothek*, published in 1821–26—a vast collection of magical texts studied as a peculiarity of past generations.[41] In the summer of 1830, the *Foreign Quarterly Review* published a comprehensive report by an anonymous writer titled "Demonology and Witchcraft," written as an elaborate review or reply to Horst's magnum opus.[42] Its author was probably George Moir (1800–1870), an advocate and writer of poetry and law who took a special interest in magic and witchcraft.[43] The article itself, which received much attention and circulated almost immediately through many popular contemporary channels, painted a picture of a deeply superstitious clergy in the late sixteenth and early seventeenth centuries. "One would almost suppose," complains the author, "that few persons at that time condescended to perform a cure by natural means."[44] Moir was also the first to question in print Aubrey's interpretation of the sign ℞. "We cannot help thinking," he remarks somewhat scornfully in a footnote, "that the prefixed characters which Ashmole interprets, to mean *Responsum Raphaelis*, seem remarkably to resemble that cabalistic looking initial which in medical prescriptions is commonly interpreted 'Recipe.'"[45] As time advanced, sympathetic accounts of Napier grew scarcer and more marginal, and his unflattering image as a medical con man, or, at the very least, the victim of dubious superstitions, became prevalent.[46]

Like that in witches and ghosts, belief in angels survived the first onslaughts of the Reformation, provoking a growing interest before an inevitable collapse. As Aubrey might have hoped, the story of the skilled doctor who built a Jacob's ladder succeeded at first in carrying Napier's practical legacy on for later generations.[47] However, in an increasingly disenchanted world, his flattering account gradually transformed into a virtual ballista in the conflict between advocates and skeptics of spiritual contact, eventually turning into a mere curiosity from a primitive era.[48] With the passage of time Lilly's biographical sketch of the doctor started to overshadow that of Aubrey, with its stress on his angelic contacts. Perhaps the collapse of belief in the possibility of communicating with angels cleared the stage for the still ongoing popular interest in practical astrology, of the kind portrayed by Lilly. Nevertheless, it was mostly the curiosities of his angelic conversations that carried the memory of Napier's skilled efforts through the first few centuries that followed his death.

Consulting an Archangel

"If we were to take seriously the historiographical injunction to work with actors' categories," suggests Joad Raymond in the introduction to his recent *Conversations with Angels*, "then we might learn by exhibiting less skepticism towards Dee's transcripts, and, if not to accept them as real, at least to investigate them as conversations."[49] As I have shown in the short historical survey above, most accounts of Napier's conferences with Raphael were used either to support the plausibility of human-angelic contacts or to denounce them as superstitions. Napier himself, however, free from the burden of future debates, was sincerely focused on their actual contribution to resolving his clients' concerns. Whether to resolve a medical case, to conclude a personal query, or as a general appeal to superior knowledge, he firmly believed he was gaining exclusive access to divine wisdom.[50] I will therefore take his records of short, spiritual dialogues with Raphael for what they really are: the transcripts of some very remarkable conversations, designed to obtain concrete information about the life and health of patients and querents.

Early modern discourse with angels is usually identified in modern scholarship with the English natural philosopher John Dee. Dee's copious transcripts of spiritual conversations, aimed mainly at gaining theological and philosophical insight, demonstrate one way by which early modern angels could assist mankind. A revival of scholarly interest around the mid-seventeenth century drew a wide range of English practitioners into attempting similar spiritual contacts, introducing a new generation of angel conjurers with similar aims.[51] Napier himself sometimes addressed Raphael with theological and ontological inquiries, but more often sought to obtain immediate practical knowledge concerning the cause, treatment, and outcome of specific medical cases, or the odds of his clients' personal prospects. This does not mean that he did not participate in the same occult tradition as Dee, yet what they shared was rather techniques than aims. Lilly maintained in his autobiography that Napier knew Dee well and that the latter was "enforced many times to sell some book or other to buy his dinner with as [sic] Dr. Napier of Linford in Buckinghamshire."[52] A mutual dinner in 1604, in which the two men discussed several alchemical works, is recorded in Napier's diary, as also is Dee's visit to Great Linford in September that year, when he just missed the rector, who was away at Luton Hoo.[53] Napier may have been therefore a "connecting link" between Dee and his mid-seventeenth-century followers, but also the voice of a largely neglected group of English angel conversants, who employed angelomancy for essentially practical purposes. These men and women, hinted at in Lilly's autobiography and in Aubrey's short surveys, mostly consulted their angels in order to gain concrete knowledge of past, present, and future events, within the context of an astrological, divinatory, or medical practice.[54] Silenced by the bias of extant documentation, memory of them might be restored through our close examination of Napier's work.

Napier's documented contacts with angels began in the spring of 1611, following a three-day absence from medical work. Prefixed by a capital *M* and penned beside the first case for which the doctor returned to take notes, an obscure piece of advice in the hand of his assistant Gerence James warns that the patient is "vngrat[eful]" and "not good to medle [with]."[55] A second prophetic statement ascribed to the archangel Michael appears after a couple of weeks of silence, this time forecasting a fruitful treatment for the forty-year-old "Mr Michels."[56] Thence, angels' prophetic verdicts begin to crowd the folios of Napier's diary.[57] In an exceptionally explicit example, he notes on the case of the eight-year-old William Walton that "M[ichael] will shew me what course I shall take with him when his father cometh."[58] Over the course of the following weeks, spiritual contacts are extended to include Raphael, Gabriel, Uriel, and Asariel, sometimes involving more than one angel in a single case.[59] By the beginning of June, Raphael's presence in the consulting room has become increasingly dominant, with all other angels virtually fading away.[60] At this point, most recorded answers are limited to disease prognosis or to forecasting the outcome of Napier's treatment.[61] Then, less than a year after having inexplicably appeared, angelic judgments become extremely infrequent. For several more years they do continue, unevenly dispersed, vanishing entirely from Napier's medical diaries by the end of 1620.[62]

However, the handful of *Interviews* that survive from the same years suggest that these changes in documentation did not result from any actual change in practice, at any rate not before the mid-1620s. In fact, comparing the names of querents in these folios with the cases recorded for the same dates in Napier's diary reveals still-astonishing proportions of angelic assistance. For instance, from the eighty-one medical cases conducted between July 26 and 28 and between August 19 and 31, 1619, as many as thirty-one (38.2 percent) appear to have been submitted to the angels for judgment.[63] Partially prefixed by the telling ℞ sign, these rough notes of spiritual interrogations unveil the overwhelming and continual role of Raphael in Napier's consulting room. Apart from an initial period of several months, Napier's spiritual conferences seem, therefore, to have been conducted almost exclusively with this particular archangel, who in the apocryphal book of Tobit had guided the young Tobias into making medicines and amulets out of the offal of a caught fish.[64] It is not clear whether it was Raphael's image as a divine healer, the ease of his summoning, or his reliable answers that guided Napier's choice of angelic aid. One way or another, it was Raphael who dominated Napier's consulting room. For ease of writing, I will treat him throughout the rest of this chapter as Napier only angel. As I will later argue, Napier saw angels as on equal theological footing with men. Yet his diaries and notes reveal a lopsided dialogue that was more of an appeal to a higher authority of knowledge than a conversation between equals. Almost four hundred years after they were put on paper, his angelic *Interviews* still project the profound awe of a man who sought divinely inspired knowledge from these wondrous beings.

Medical Cases

Five weeks ill with "dysenteria" and voiding blood, Michael Horwood, a young man from Newport, came to seek Napier's help in August 1620. In his medical diary, the doctor noted personal details and physical complaints, and erected an astrological chart for the time of visit. To the left of the chart, he recorded a short prognosis: "℞ he is likely to die with out recovery" (see fig. 6).⁶⁵ Notwithstanding the harsh judgment, he prescribed the patient a list of herbal remedies. George Katlen, a mad middle-aged man from a village near Luton Hoo, who would at times "pisse in the chimney corner," was brought to Great Linford a couple of months later.⁶⁶ Napier gave him two Jovian sigils and, in a different ink (i.e., on a later occasion), noted just above the case notes: "℞ he will mend by gods grace." Short prognostic verdicts of this kind, most of them prefixed by the revealing mark for the archangel's name, are scattered throughout Napier's medical diaries in varied proportions, encompassing in a handful of words the good or bad fortune of the anxious patient.

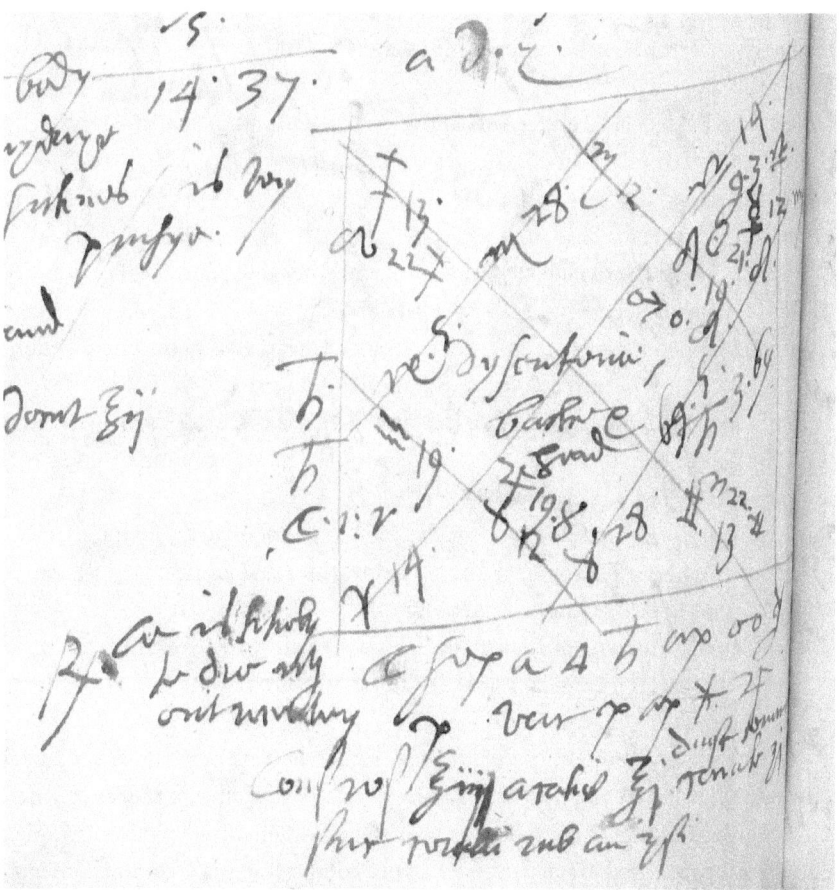

Fig. 6 A short disease prognosis attributed to the archangel Raphael, in Richard Napier's medical diaries. The note reads, "℞ he is likely / to die with / out recovery." Ashm. 414, fol. 134v (Horwood). By permission of the Bodleian Library, University of Oxford.

Providing disease prognosis was a common practice in early modern medicine. Reassuringly answered, it could dispel the greatest anxiety of a sick person in an era with an abundance of widespread fatal ailments: will I get better and when? For the seventeenth-century patient, this was an important piece of information he expected to receive from his medical caregiver.[67] The content and wording of Napier's angelic prognoses appear in an assortment of forms. Whether laconic, as in the couple of previous examples, or more generous in details, as in others, they always contained a determinate, concrete verdict of either life or death. In the case of one "very shortwynded" middle-aged woman, for example, Napier recorded an exceptionally long list of symptoms and a detailed medical history, but erected no astrological chart. He noted, however, that "she will dye about halfe a yere sinc by the consumption of her lengs [lungs] & liver."[68] Mrs. Dacres, a seventy-seven-year-old woman from the nearby village of Wolverton, was afforded a more cheering reply: "℞ she will amend by gods grace."[69]

An even better sense of patients' mortal anxieties and the nature of Napier's dialogues with the archangel emerges from the *Interviews*:

- Q: "Nicholas Halfehead, 6 y[ears], of Hartfordshire, troubled with the falling sickness & lamenes, halfe a yere, wheather to be cured & how." ℞: "he shall mend."[70]
- Q: "A mad woman that would not come near me, dwelling beyond London, wheather to be cured & how." ℞: "shall not be cured."[71]
- Q: "a scottish fellow sent by Mr Jerryin to be cured of madnes by fits, useth mutch to cough & smyle, a mery fellow." ℞: "he wilbe cured by gods grace."[72]
- Q: "Goody Shepherd Redburne, lame in her syde a longe tyme, [pain?] running vp & downe, can find no helpe, the cause & the cure." ℞: "neaver to be cured, the finger of god."[73]
- Q: "Georg Kathelyne of Dunstable, no vomyt will worke with him or cure his disease, the cause & cure." ℞: "not to be cured."[74]

The format and content of the *Interviews*, matched against Napier's medical diaries, suggest that Raphael was not invoked upon each separate encounter. Gaining angelic attention, as I will later delineate, involved a fairly complex ritual with prolonged prayers, and therefore conducting the ceremony for each individual patient simply did not make sense. Those distressed patients, who sometimes traveled great distances to Great Linford and lodged in the village often had to wait a few days to acquire their angelic verdict. Keeping this time lag in mind, comparing the two texts can provide a clearer idea of how a querent's distress was translated into the language of angelic communication. A typical case is that of Sir Richard Spencer (d. 1624) of the Hertfordshire village of Great Offley. In a medical note from August 5, 1620, Napier erected an astrological chart for the time of visit, recorded complaint and prescription, and concluded with the following prognosis: "will live halfe a yeere neere [nearly half a year] & will live [sic] about a month before it, neere to Easter."[75] The *Interviews* disclose Raphael's correlate prophecy, the actual source for Napier's harsh verdict: "Sir Rich[ard] Spencer will live about halfe a yere & die of a consumption."[76] In a later interview from August 26, Raphael declared likewise that Sir Richard "will recover for a while, but will dye in the end, a fortnight before Easter next," and Napier went back to his former notes to append the new information.[77]

The day before the nobleman's visit, Napier received for inspection the urine of Thomas Hollys (or Hollis) from Hanslope: "a thick white puddle water, full of thick shyny matter mixed with blood."[78] Beside the astrological chart he noted the patient's complaint: "back mutch payned, maketh water often & with great heate & payne, brought to great extremity & poverty." His appeal to the archangel, recorded among the *Interviews* for the same day, inquired "wheather poor old Thomas Hollys of Hanslop be ever to be cured of that shynye stuffe mixed with blood in his yard [penis] pissing, oftene & but litle, misrable, payned in his backe & making of vryne & will returne shortly into a consumption."[79] "To be feared," was Raphael's ominous reply.[80] Here, the query to the archangel simply repeated the information recorded in the medical note. In other cases, the questions recorded in the *Interviews* shed further light on the information that the diaries disclose: Mary Kings, a thirty-one-year-old woman who could not "abyde her husb[and] nor her child,"[81] was prescribed two Jovian sigils. The correlating query in the *Interviews* reports, Q: "Mary Kings of Hamfford, distracted." R: "she will amend."[82] Mary Whitlerke, a mad woman who was ripping her clothes in pieces, "bound in her bed with cords at night & at day tyme is chayned at a post,"[83] was prescribed mercurial tabs and a couple of Jovian sigils. Napier's query to the archangel reveals her family's further concerns: Q: "diseased, will seeke to hang her selfe." R: "she will mend."[84]

Perhaps the most intriguing practice associated with Napier's reputation as a magus is his alleged appeal to angelic assistance in determining the course of cure. Such specific spiritual queries, along with inquiries regarding the cause of illness, position Napier's angelic conversations at the core of his medical practice rather than as a mere sideline and curiosity. In the case of Sir William Andrews, who "fell into a shaking fit" and then "vomited strongly," Napier appended the prescription of treacle water with cinnamon and a cordial drink, with an instruction by Raphael to let the patient blood (see fig. 7, and the more

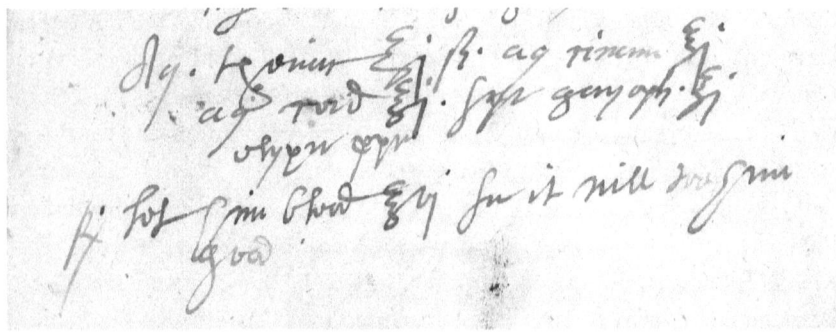

Fig. 7 Richard Napier's prescription in the case of Sir William Andrews, March 31, 1618. The indicative sign for Raphael's response precedes an instruction to let the patient two ounces of blood, promising that "it will doe him good." Ashm. 201, fol. 43v (Andrews). By permission of the Bodleian Library, University of Oxford.

detailed angelic prescription in fig. 8).[85] Similar examples are scattered throughout his diaries, albeit quite rarely.[86] In the *Interviews*, inquiries about the course of cure constitute only a minority of the items, yet examples are more than a handful, and it is reasonable to assume that they represent a much wider appeal to Raphael in such matters. In the absence of any other firsthand testimonies for the involvement of angels in early modern medicine, Napier's diaries and notes provide an extremely rare glimpse into this extraordinary practice. The archangel's prescriptions and medical directives, perhaps the prime products of Napier's spiritual dialogues, remain an intrinsic part of his professional legacy and reveal the extent of Raphael's contribution to the resolution of specific medical cases in his consulting room.

> Q: "Wheather Mr Vnderhill will be recovered of his ey[e] sight & hearing, how & by what meanes." ℞: "a purg, dyet drinke & the oyntment."[87]
>
> Q: "James Tickle of Knebworte hath a hard knot growing with in the Inside of his belly, not to be seen, wheather & how to be cured." ℞: "a good strong dyet drinke will helpe."[88]
>
> Q: "Elis Wilshew, a bloody water 3 weekes & very weak, wheather to be cured & the cause." ℞: "to be cured with rest, gent[le] things."[89]
>
> Q: "What course to be taken with Sibill who is mopish, so hath bene 6 years vppon greife taken for her father." ℞: "Purg her melancholy lusely & give her a S[igil of] ♃ & S[igil of] ♃ in *aq[ua] hyp[pericum]*."[90]
>
> Q: "What I shall vse to my swelled ankle, wheather myne owne vryne with ashes sifted in it, or with a litle black sope [soup]."[91]

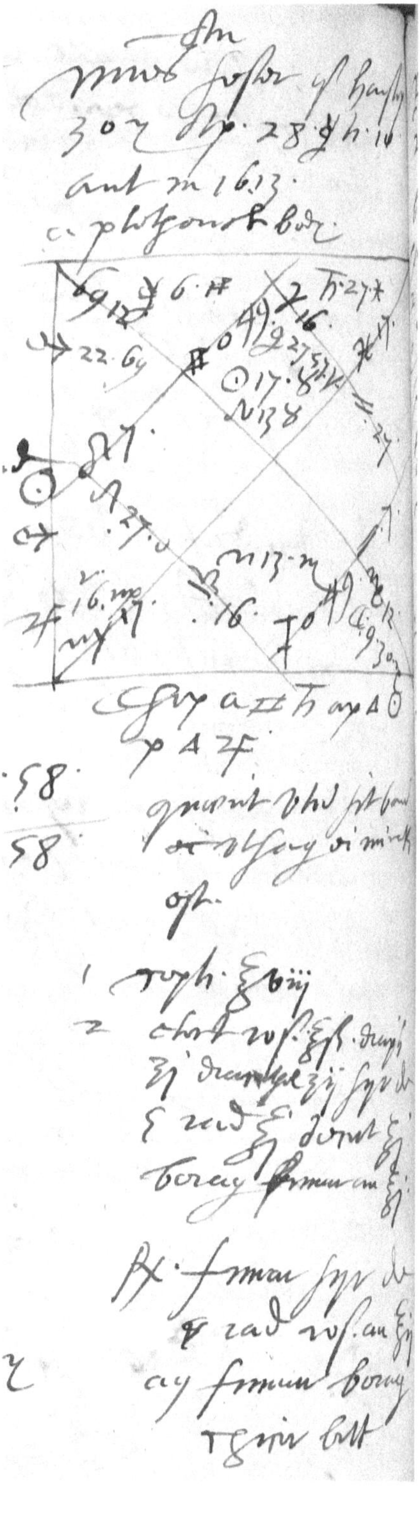

Fig. 8 Richard Napier's medical notes in the case of An Foster, April 28, 1613, ending with a prescription attributed to the archangel Raphael: "℞ *fumar[iæ], syr[up] de / rad[icibus], ros[arum] an[a]* ℥ij [each two ounces] / *aq[uarum] fumar[iæ] borag[inis] / chicor[ium]* bitt[er]." Ashm. 199, fol. 106v (Foster). By permission of the Bodleian Library, University of Oxford.

Some prescriptions were much more exhaustive. For "old Mrs Pots," for instance, Raphael recommended the following: "[C]lyster her & give her pils, & then spring & fall give her the long blysering playster, keepe it open 3 weekes & then apply the playster of london triacle adding choles, ginger, pepper, pellitory, white amber & oyle of euph[orbia] & castered. let her take [gaules?] before the blyst[ering] playster & take the leaches the 4 quarters of the yere."[92] Yet the archangel was not always cooperative, or even correct. In the case of Mrs. Panton of London, who came to see Napier "for her paynfull swelling, red at [nose?] & cough,"[93] Napier consulted his angel on the very same day. "Wheather Mrs Panton shall be cured of her consumpt[ion]," he inquired, "what course best, [and] wheather bewitched." Raphael curiously replied that "she will mary," perhaps as a token of her returning to good health. "Not soe," Napier later noted, "she dyed of consu[mption]."[94] The archangel was also invoked several times regarding the best remedy for Napier's cousin Elizabeth Marsh. "What that speciall thing is, that will helpe my cosen Eli[zabeth] Marsh of the mother & free her from it," he inquired almost desperately in the summer of 1619, but received no answer.[95] This was also the case with Susan Feste of Newport, who was "payned in the share or back."[96] "Which physick," inquired Napier, but the archangel was silent.

Another recurring query concerned the nature of the medical ailment. As I have shown in the chapter on astrological medicine, Napier mainly extracted the cause, which defined the essence of the disease, from certain celestial constellations in the astrological chart. From the early 1610s on, he also attempted supernatural help, at least in some of his more difficult cases. "[T]he cause of Mrs Byrds disease," Napier queried Raphael in a typical plea; "corupt blood, let her blood," was the archangel's reply.[97] Sir Thomas Temple's wife was deeply disturbed as her husband repeatedly woke up crying in the wee hours of the night. Napier consulted Raphael, who ruled that "the cause is no temptation of sathan to evill but his owne tender conscienc touched with the remorse of his sins which move him in a manner to despy[re] of his salvation."[98] Clearly, the archangel had full knowledge of all physical ailments, and Napier occasionally shared his hesitations regarding a medical judgment or, in the more tricky cases, presented an "open question":

Q: "Dorothy Parot of Newp[ort], wheather the pox." ℞: "it is, will amend."[99]

Q: "Wheather Mr Latch . . . being troubled with epilept[ic] convult[ion] in his sleepe & at no tyme els, be [sick] by witches or els by stony impostium of the fantasy."[100]

Q: "Johne Midle of Cranfild, 15 y[ears], taken with a stony stifnes in his eyes, shooting in boath eyes, making him blynd for the tyme. wheather naturall & curable & how." ℞: "witchery, a sigill will helpe."¹⁰¹

Even during the high days of Napier's angelic divination, referring to angelic help was only conducted selectively. Within these instances, a significant portion concerned suspected bewitchment or madness. In fact, out of a total of 120 surveyed medical queries in the *Interviews*, 25 (20.8 percent) can be identified as such: more than four times their general proportion in Napier's medical practice.¹⁰² This statistical significance should be attributed, I believe, to the greater complexity of mental cases, or to the special knowledge ascribed to spiritual beings regarding preternatural activities. Alice Tompson, a "distracted" young woman who "was mended & after relapsed,"¹⁰³ came in July 1619 to see Napier, who immediately suspected bewitchment. This time Raphael failed to supply an answer, but the doctor himself later noted that "she was" indeed bewitched. In a reply to a query made on the same day for Elizabeth Shilton, however, Raphael declared the poor woman to be incurably bewitched.¹⁰⁴ The archangel could also confirm or deny the identity of the suspected witch, and Napier sometimes made use of his ability to acquire such valuable information. It is not clear, though, whether he later made any attempt to denounce the accused.

Q: "Rebecca Wallee of Westmyster, full of fancyes, present now at [Bedford?], accuseth on[e] Chrysty of witchery, an honest woman full of talke."¹⁰⁵

Q: "Wheather Mrs Goodwyne of Cardmaston be illspoken & beworded by Goody miller or els by the wife of on[e] Stamton." ℞: "she is, on[e] horse leach, Sig[il] & a Sig[il] [sic]."¹⁰⁶

Q: "Wheather Katherine Sanders *sit malefica et* Eliz[abeth] Ballma."¹⁰⁷

Suspected pregnancy also provoked occasional appeals to Raphael. Goody Stuchbury, a forty-two-year-old woman from Winslow, came to Napier, being "swelled in her chest & belly, burneth vryne, anguish & red."¹⁰⁸ Guided by these symptoms, he inquired of Raphael whether she was pregnant and noted a positive answer.¹⁰⁹ Napier's cousin Lady Rotheram also turned out to be with child, according to the archangel,¹¹⁰ and a similar reply was given for his patron's wife, Lady Wentworth, yet Napier later noted that "she was not."¹¹¹ Sir Thomas Myddleton, the husband of Napier's sister Mary, suspected that his wife was pregnant, but "feareth abortion."¹¹² The doctor erected an astrological chart that confirmed the pregnancy, but also consulted Raphael, who provided the same answer.[113]

Other queries were devoted to the forecasted success or failure of Napier's treatment. Attentive to the outcome of his medical orders, Napier sometimes appealed to the archangel to double-check his prescribed cure. "Wheather my physick will help Elis Ashwell of langport who is mopish,"[114] he asked Raphael concerning one troubled patient. "Wheather old Mrs Booth of the bay shall receave good by the sigil & purg which I sent her,"[115] he inquired regarding another. Of the latter, the archangel answered rather generally that the prescribed sigil was "good agaynst devilishal fears & suspicions," and Napier could thus substantiate its benefits. Although he was less concerned than was Forman with his personal and professional image as a successful physician, Napier was understandably reluctant to take on a hopeless case and often consulted Raphael when he anticipated failure.[116] "Wheather if I take my lord of Ruthen & heyre, The lord Reste, I shall cure him,"[117] he asked Raphael in the summer of 1619. "He will never be cured quite but I shall doe him mutch good," he concluded contentedly from the archangel's reply. "Wheather I shall help Sir Arthur Harys her [sic] sister of her melancholy [and] madnes,"[118] he inquired elsewhere, noting beside Raphael's reassuring answer: "I shall have helpen her."

- Q: "my L[ord] Purb[eck], his vomyt, wheather it will doe him good, wheather the bones will mend."[119]
- Q: "Wheather Staphen Vle of Cranfye [Cranfield?], 70 y[ears], now very vnruly & mild, will [be] cured with the conse[rve] that I have prescribed."[120]
- Q: "Wheather Mrs Mary Archer have not the grene sicknes ... [and] wheather I shall cure her & how." R: "she shallbe cured by gods grace."[121]
- Q: "Wheather Mr Page of All soules [that] was crazed with study, willbe presently recovered by the physick I send him." R: "wyll by fits. he will doe partly well yet will be ill by fits & not perfectedly cured."[122]
- Q: "Wheather old Sir Will[iam] Andrews will receave good by the purg & horse leaches, of his melancholy & payne about the pit of his stomach."[123]

Personal Prospects

Questions regarding personal concerns or aspirations constitute about half of the surviving *Interviews*, scattered among the medical queries that I have just surveyed. Yet, unlike the latter, almost none of these inquiries regarding the major cornerstones of life and death are reflected in Napier's consultation notes.[124] Thus, although they appear, at first glance, as typical divinatory questions, it is possible that they represent an activity that was considerably different from standard early modern divination. In the following section, I will examine the various issues with which these queries dealt, but also attempt to resolve this apparent discrepancy between the *Interviews* and the diaries.

Modern accounts of angelomancy, based on medieval and early modern literature, usually attach this practice to the recovery of lost goods.[125] Napier's *Interviews*, however, mostly touched upon the more basic concerns of life expectancy, marital prospects, future children, and heirs: roughly the same issues that recur in other modes of premodern divination, such as astrology, geomancy, numerology, and necromancy.[126] Therefore, rather than highlight Napier's spiritual conversations as a distinct method of foresight, they seem to simply reflect his clients' foremost anxieties. The fundamental role of early modern divination, argues Keith Thomas in *Religion and the Decline of Magic*, "was to shift the responsibility away from the actor, to provide him with a justification for taking a leap in the dark, and to screw him up into making a decision whose outcome was unpredictable by normal means."[127] Dealing with essentially unchangeable life events such as marriage, offspring, and death, only a small minority of Napier's queries fit this portrayal. This does not point, I suggest, to the distinct character of Napier's particular practice, but rather to a needed amendment to Thomas's claim.

A significant share of the personal, nonmedical queries concerned life expectancy.[128] As one reads through the *Interviews*, they frequently appear as straightforward judgments, such as "Old Sir Will[iam] Andrews [shall] live about 6 y[ears] & then die."[129] Only the differences in ink between the sentence's prefix (sometimes only the subject's name) and its prophetic ending reveal that the archangel's answers were later additions to pre-prepared lists of questions, appended perhaps during the ceremony itself or immediately after. Some folios are composed almost exclusively of such longevity questions, several of which appear repetitively (usually twice or thrice) within the same sheet.[130] Their considerable proportion in Napier's *Interviews*, and the fact that they relate, by and large, to members of the gentry, seem to testify to the insecure feelings of the higher levels of English society concerning the extent of their lives.[131]

> Q: "How long old Sir Griffayn will live & what good he will fynd by his physick." ℞: "will live 4 years longer."[132]
> Q: "When Sir Will[iam] Andrews will die & of what disease." ℞: "he will live better then 6 yeres."[133]
> Q: "How long Sir Th[omas] mydleton shall live & wheather he shall die of a consumption or of the stone or of greefe."[134]

As these last examples disclose, some of the queries followed a specific medical condition that either Napier or the patient saw as life threatening. Others were probably aimed at easing a general anxiety over death.

A singular set of questions was intended to sort out which one of a couple would live longer: the husband or wife. Thomas has already noted that these queries were fairly common in seventeenth-century divination, and rather surprisingly has linked them to an apparent death wish for one's spouse.[135] In an era when a second and even a third marriage after being widowed was extremely common, it is more likely, however, that they were meant to convey the future prosperity of the querent (that is, if he or she would be the one to survive the present marriage).[136]

> Q: "wheather Mr Th[omas] dighby shall bury his wife [and] yf so how long henc & in what prosperity." R: "he shall live & mary agayne & have a good wife."[137]
> Q: "L[ord] Wyntwort [Wentworth]" R: "will live some 8 y[ears] [and] die of a bing feaver. his ladye will live long & mary another."[138]
> Q: "Sir Fr[ancis] Fortiscue" R: "will live long, but his wife will not live but a few years, as 4 or 6 years."[139]
> Q: "Wheather Mr Paller will [out]live his wife & doe any thing for her." R: "she shall out live him & fynd favour at his hands."[140]

A related issue of concern for Napier's querents was marriage.[141] While some of the questions on this subject were fairly straightforward, others arose from a general hope of well-being, depicting early modern marriage as a token of health and prosperity. In an example of the latter, Napier inquired, "wheather ms Frances Digby shall [be] recovered of her melancholy & mary a good fortunat husb[and]."[142] The girl's marriage was evidently seen as a mere by-product of recovery and return to regular life. Other cases voiced more specific expectations, such as Lady Adimer's query whether a certain "Mr Stealus" shall win her daughter's love and "come to good preferment."[143] Another curious query recorded an anxious mother who inquired if her son should marry his mother's "poore mayd" or "continue still foolish & idle headed for lack of on[e]."[144] "Purge him well with vomyts or purges or clysters or all & leaches," the archangel advised, "[and] let him marry wher [h]is mynd is set." Marriage was an important sign of social status, and some were apparently more fortunate than others. Sir Arthur Throckmorton, inquiring whether his youngest daughter shall be married to a "knight or lord," learned to his content that she will marry "a very rich knight & shall have very good fortune & live quetly [quietly] & merrily."[145]

Q: "Wheather Mr John Monesee shall mary & what dowery & portion of what degree, & how affected, & who shall live longest, & wheather a good match." ℞: "a yere & a halfe hence betwixt July 23 & miklmas [Michaelmas], on[e] of good birth & ritches."[146]

Q: "Wheather Rob[ert] Nap[ier] shall mary Complers daughter or no[t], or with whom, & wheather he shall have the like dowery & portion & shalbe a good match." ℞: "he shall with mutch adoe."[147]

"How long my lord Ruthen shall live & who shall succeede him,"[148] Napier asked Raphael in 1619, voicing an eternal concern over children and heirs. "The next child a wench, but [he] shall have a sone," the archangel assured him. Another gentleman inquired whether he will entail his land to his daughters, or rather remarry and leave it to a future son, "if ever he have any."[149] "He will entayle it to his wenches," replied Raphael, "& will not [out]live this wife & neaver after will mary agayne." "Having issue," the early modern term for producing a legitimate heir, encompassed broad concerns, especially among the gentry, regarding self-esteem, family continuance, and the fate of family title and property. Napier's cousin-in-law Thomas Marsh, for example, inquired whether his son of the same name "shall live long & have a happy mariadge & possesse his fathers lands."[150] A rather illuminating case, reinforcing the sentient nature of the interlocutor of Napier's *Interviews*, concerned Napier's eldest nephew, Sir Robert. "Rob[ert] Napeyr shall live 40 y[ears] & have 3 wenches," proclaimed an early prophecy, "[and] the living is likely in the end to desc[end] to Rich[ard]."[151] Yet in a different ink, Napier later remarked that, "sinc ℞ hath denied it & affir[meth] he shall have sons to inherit."[152] Obviously, the archangel did not hesitate to change his mind, leaving young Robert with a more satisfying prospect.[153]

Q: "Wh[eather] Sir Jane shall out live his brother & obtyne his lands, he or his children." ℞: "he shall out live him & the lands shall come to him."[154]

Q: "Wheather he, Mr Adolph Andrews, will increate his wife agayne & lay with her & have a sone by her & wheather that sone shall inherit his lands." ℞: "he will not recreate her, yet will keepe the child, & will greeve & die of a consumption."[155]

Angelomancy could also unravel genuine personal doubts, as in the case of Sir James Evington, another of Napier's nephews. The son of Sir Francis Evington, a merchant tailor and alderman of London, and Napier's sister Margaret, James was pretty much addicted to horse racing and, from the late 1610s, had started to experience considerable financial difficulty. In a letter dated 1618 he made a first attempt at deploying his uncle's extraordinary skills, inquiring whether he was likely to win the cup at the forthcoming Stamford races and whether he should afterward sell his racehorse and the remainder of his lands in Bedfordshire.[156] A couple of years later, as his finances became worse, he came to Great Linford to inquire about the possibility of selling Casewick Hall, a medieval manor house in Lincolnshire that was his main family property and place of residence.[157] Napier used his astrological skill to give judgment, but also consulted Raphael "wheather Sir James shall quietly & peacleablely [peaceably] posses the lands his father left him, . . . [and] wheather better to keepe them [or] to sell his land."[158] The diaries record a second encounter, five days later, in which the same question was posed.[159] Under the astrological chart, Napier noted several planetary aspects, made the necessary calculations, and declared the lands should not be sold. Once again, he also consulted his angel, providing a fuller account of the apparently complicated situation: "Sir James is come of purpose to know, sithenc his lawyers do signify that his state is not fully formed, from his sone & heyre of whose hayre it was purchased, his sone being lend when he was vnder years, & his father decayed. [The] question is wheather it were not better to sell it, as he thinketh that he can, with sutch assuranc as he hath & no better, & to imply the way eyther in vsing or els to buye an offic."[160] Sir James demanded further to know whether he would outlive his wife and marry another, "very ritch."[161] The archangel gave a disappointing reply, and the knight left Great Linford on the following day, in heavy rain. "God be his good sheperd," Napier recorded his silent plea.[162] In this case, therefore, as in Napier's medical queries to the archangel, the regular case notes and the *Interviews* complete each other's stories. They also point to astrology and angelomancy as alternative yet complementary means of divining a course of action in personal dilemmas.

Sir James Evington's case is perhaps an outstanding example, but it was not the only angelic advice in Napier's consulting room to provoke action. Constituting just a small fraction of the surviving *Interviews*, such inquiries concerned hiring servants and keepers, dubious real estate transactions, risky lawsuits, and more.[163] Some of the most intriguing queries regarded the efficacy of Napier's own pursuits, fine-tuning his future professional efforts. "Wheather the write me write vppon will prove right," he asked the archangel regarding an unspecified paper, in the same way that he questioned elsewhere the validity of others' alchemical and medical works.[164] In a more enigmatic inquiry, he demanded to know whether and when he "shal obtayne a S [Seer's stone?]."[165] This seems to resemble Ashmole's use of astrology to find out if he should ever succeed in producing the "magic specula" of Paracelsus.[166] Other examples concerned some of his textual tools of healing:

Q: "Wheather the manuscript agaynst witch[ery] & vppon the divell be suff[icient] for that purpose, if it should be vsed. a frynd [friend] gave it that did helpe many." R: "[It] is a good booke vsed with fayth & good devotion, the body well purged & a sigill."[167]

Q: "My prayer agaynst w[itchery] & sorc[ery] in the end of my booke." R: "very good."[168]

Most of Napier's self-queries provided, even if obliquely, a "call to action," and so too did some of the appeals posed on behalf of his relatives and acquaintances. These latter, however, were marginal among his nonmedical spiritual inquiries, about half of which concerned members of the gentry. As I have shown, most of these queries did not arise from specific personal dilemmas but rather concerned the prospects of unchangeable life events. Only rarely were they reflected in Napier's regular diaries or did they even follow any recorded consultation by the same querent. Whether they were posed by clients visiting Great Linford in person, submitted remotely through a messenger, or initiated by Napier himself is unclear. Judging by their dispersal in the *Interviews* and by the frequent clusters of family queries,[169] it seems that at least some of them did not originate from standard medical encounters in Napier's consulting room. Instead, they appear more to resemble his astrological and medical correspondence, in which the questions of colleagues, friends, and family were reviewed away from the routine of everyday practice.[170]

"An angell's sight"

> The Magicians, now, use a Crystal-Sphere, or Mineral-Pearl for this purpose, which is inspected by a Boy, or sometimes by the Querent himself. There are certain Formula's of Prayer to be used before they make the inspection, which they term a Call. In a Manuscript of Dr. *Forman* of *Lambeth* (which Mr. *Elias Ashmole* had) is a Discourse of this, and the Prayer. Also there is the Call which Dr. *Nepier* did use.[171]

The idea of angels as transmitters of divine knowledge to able magi through visual or oral apparitions has its roots in both biblical tales and early Christian culture. The Hebraic tradition offered a complex hierarchy of angels who had diverse functions, ranks, and importance. Arabic and Greek ideas and texts and practices of spiritual magic all fused in the later Middle Ages with Christian liturgy and ritual practice, especially exorcism, initiating a practice that was designed to summon angels proactively through the power of ceremony and prayer.[172] Renaissance thinkers such as Marsilio Ficino, Cornelius Agrippa, and Pico della Mirandola formulated these ideas in a distinct fashion, weaving into them Neoplatonic notions and elements from the Jewish Kabbala.[173] They believed that divine influence descended from the world of spirits to the stellar world, which in turn governed terrestrial objects, including human bodies.[174] Angels thus influenced the movements of planets and stars, or could at least interpret them.[175] In the previous chapter, I showed how Napier's astral magic was inspired by his willingness to act proactively in a world governed by celestial forces. His angelic magic could likewise be viewed as a practical aspect of the same worldview, but one involving knowledge acquisition rather than purposeful action. Thus seen, Napier's conversations with angels come into view as an extension of his astrological-medical practice, in which Raphael, like the planets and stars, was an informant who revealed obscure knowledge.

Several theoretical aspects of angelic magic were bound up in an astrological context, and many of its practices were informed by astrology or involved astrological conditions. According to *Sepher Raziel*, a classical medieval manual, directions for invoking angels varied according to season, time of day, and purpose of the session.[176] These might not have been originally designed as astrological guidelines, but for seers of the seventeenth century, many of whom were practicing astrologers, such directives were an obvious and expected aspect of the angelic art. Ashmole himself offered explicit astrological rules for angelic invocation, asserting that one could work only on even days in the "increasing hours" (from sunrise to noon and from sunset to midnight) and when the Sun was "well placed with a beneficent planet reigning."[177] Moreover, the success of the "Call," the actual procedure for summoning angels, was dependent upon several factors, such as the content and form of prayer and the purity of the conjurer's life and the time of operation, in much the same way as the success of astral magic depended upon the correspondence between material, image, and timing.

Invoking a "Call"

> I charg you by all the firmaments of heaven & by all the blessed Angels & arkangles & by all the holy saincts, martyrs [and] confessors, that presently you appear in this Christall & that you show your selfs vnto N[ame][178] & that you show & reveale vnto him or her all sutch things as I shall demand of you truely and [with] sincerity, without falsehood or

dissimulation, as you will answere it at the dreadfull day of iudgement before the high tribunall of our omnipotent lord Jehovah fiat. Then say the creede, then say the call agayne, then the *patre n[ostre] & av[e marya]* & then the call agayn & then the creede, *patre n[ostre] & a[ve marya]*, vntill you have sayed it 5 times over. And betwixed every question say a *patre n[ostre] & Ave mary[a]* & the chardge.[179]

Texts detailing invocation procedures that circulated in manuscript form among clerics and practitioners of the Middle Ages, reinforced by the Renaissance revival of Christian magic, disseminated practical instructions for conjuring angels and spirits throughout Western Christendom (see fig. 9). These manuals incorporated a wide range of ritual practices from the Christian liturgy, designed to provide the magus with spiritual and intellectual powers over spirits and demons.[180] Forman, himself a conjurer of spirits and angels, possessed copies of the *Ars Notoria*, *Sepher Raziel*, *The Picatrix*, and Agrippa's *De Occulta Philosophia*, to which Napier added his own copy of *Heptameron* (falsely attributed to Pietro d'Abano).[181] A legitimate heir to this long-established textual tradition, Napier's practical methods of invocation and scrying, we may reasonably assume, rested upon texts that derived from this pre-Reformation religious and magical milieu.

For a Christian magus such as Napier, the act of summoning an angel was a religious rite, in which rigorous prayer played an essential part.[182] The above text of invocation, one of about half a dozen different versions that survive in his archive, may attest, at least to some extent, to his actual modus operandi in performing the complex ritual.[183] The recital of Catholic formulae such as the Lord's Prayer, creed, and Hail Mary, as well as the request for true answers from chaste spirits, places the devout rector within the mainstream of angelic magic traditions.[184] Rigorous litany stood at the core of any invocation procedure, and some rituals were made in an almost continuous state of prayer. Copious repetition of certain expressions was also common and contributed to the ecstatic state of mind of both operator and medium. Summoning texts were sometimes assembled as an appeal to the desired angels, but more often designed as a plea to God himself. Such a direct petition to the divine "to send his angels to illuminate the crystal" not only ensured that the discourse would not be misleading, but also emphasized the ultimate target of the ritual.[185]

Skeptics of angelic apparitions usually argued that conjurers could be easily fooled by demons in disguise, substituting false for true knowledge.[186] It was on these grounds that Méric Casaubon attacked the efforts of John Dee when he published his spiritual diaries in 1659, and that John Prideaux, Napier's client in angelic divination, sent his warning to "popish votaries" and enthusiasts,

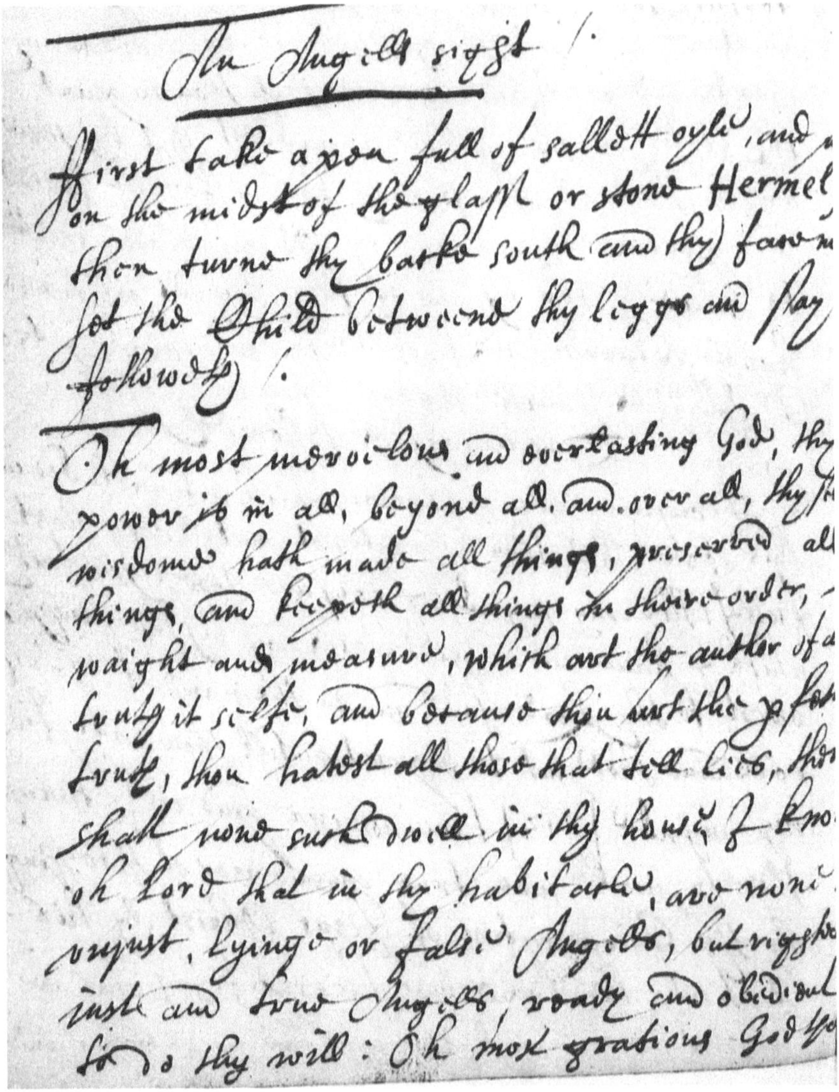

An Angells sight /.
ffirst take a spoon full of sallett oyle, and …
on the midst of the glass or stone Hermet…
then turne thy backe south and thy face …
set the Child betweene thy leggs and say …
followeth /.
Oh most mervailous and everlasting God, thy
power is in all, beyond all, and over all, thy …
wisdome hath made all things, preserved all
things, and keepeth all things in there order, …
waight and measure, which art the author of …
truth it selfe, and because thou art the …
truth, thou hatest all those that tell lies, there
shall none such dwell in thy house I kno…
oh Lord that in thy habitacle, are none
unjust, lying or false Angells, but right…
just and true Angells, ready and obedient
to doe thy will: Oh most gratious God …

Fig. 9 A page from Elias Ashmole's instructions for summoning angels, titled "An Angells sight." © The British Library Board, Sloane 3846, fol. 66v.

"least in their unwarrantable devotions, instead of an Angel of light, they meet sometime with a worse commodity."[187] The English bishop and controversialist Joseph Hall argued in his 1654 *Cases of Conscience* that angels were not likely to be commanded by "ignorant, or vicious persons," and warned that "the Devill hath an unseen hand in these effects, which hee marvailously brings about."[188] Large segments of many invocation texts were dedicated accordingly to an appeal to God (or to Christ) to deter Satan from sending his fraudulent demons in the disguise of true angels. Demonic epistemological powers were themselves considerable, but angels' knowledge was more certain because their ultimate source of revelation was God. Moreover, conjuring evil spirits was utterly unlawful. A large cadre of scholars tried therefore to provide guidance for discerning genuine angelic contacts from demonic delusions, producing numerous versions of appeals as in the examples above.[189]

Spiritual magic was based on the notion that one could reach God through a variety of essentially religious exercises, such as prayer, fasting, and contemplation. Rather than the mechanical operation of a set formula, rituals for summoning angels constituted the operators' sincere appeal to God to extend his privilege of divine knowledge.[190] Unlike most kinds of ritual magic, angelomancy was thus a fundamentally interiorized tradition, concerned with direct experience of the deity through an ascent up the scale of creation. Both Agrippa and Dee argued that the magus could gain an understanding of the cosmos not only through learning or access to knowledgeable intermediaries, but also through direct intellectual contact with the divine, which was the true aim of angelic conferences.[191]

The Mechanics of Angelic Conversations

As with nearly all sorts of image and ritual magic, preparation of body, mind, and soul was necessary before the onset of each angelic invocation. This must have been all the more true for procedures that were aimed at experiencing the Divinity himself. The seeker of angels was required to lead a pure life, follow a good diet, and exercise in order to interact with angelic spirits, which were attracted to the purity of the soul.[192] In a short sketch of Dee's angelic legacy, Ashmole outlined some of the necessary preparations: the magus must "3 days before abstaine from Coitus, & Gluttony &c," and should "wash hands, face, cut nails, shave the beard, wash all."[193] Lilly remarked that certain spirits loved "neatness and cleanliness in apparel, a strict diet, an upright life, [and] fervent prayers unto God."[194] His account of the invocations conducted by John Evans described how the latter had regularly read his Book of Common Prayer, worn the surplice, and lived "orderly" for a fortnight until the angel appeared.[195] Of at least equal importance was the purity of the scryer, the human medium who directly communicated with the angel on behalf of the magus.

Employing a young boy or girl as a medium for angelic spirits was a technique of medieval origin that was still advocated by early modern texts of invocation.[196] "When a spirit is raised," claimed one Southampton magus in 1631, "none hath power to see it but children of eleven and twelve years of age, or such as are true maids."[197] Two hundred years earlier, the Norfolk scribe and reeve Robert Reynys described a ritual in which the conjurer took a child between the ages of seven and thirteen, placed him between his legs, and wrote on his cleaned thumbnail the name "AGLA," upon which three angels appeared on the boy's nail.[198] Most reported angelic conferences from the late sixteenth and seventeenth centuries were achieved nonetheless through an adult scryer or without mediation at all.[199] Dee's mediums were homogenously adult males (except for a short period in which he employed his own son Arthur). Lilly reported a woman by the name of Sarah Skelhorn, who was a "speculatrix unto one *Arthur Gauntlet* . . . professing Physick."[200] "This *Sarah*," he observed, "had a perfect sight, and indeed the best eyes for that purpose." Forman, who employed John Goodage and Stephen Mitchel as scryers, noted that a scryer's body must be "purged from evil humours."[201] Operator and medium could nevertheless be the same person, and scryers are virtually absent from Aubrey's descriptions of spiritual contacts.[202] Napier's diaries lack any mention of a scryer, but nor do they suggest he saw or heard his angels himself. The most revealing testimony on this matter is perhaps that of Edward Waller, quoted above by Aubrey, who as a child had witnessed the doctor praying by himself in his gallery, returning with the archangel's prognosis and course of cure.

One of Aubrey's intriguing anecdotes in the *Miscellanies* concerned a red "Consecrated Berill," originally possessed by a certain Norfolk minister.[203] The stone, with its text of invocation, was later passed on to the minister's friend, a miller, who could also see angelic visions in it. "[B]oth did work great Cures with it," reports Aubrey, "and in the *Berill* they did see, either the Receipt in Writing, or else the Herb."[204] The beryl then came into the hands of a London gentleman, "who did tell strange things by it," leading to its confiscation by the city authorities. It eventually ended up with Sir Edward Harley of Herefordshire, where Aubrey (so he tells us) had seen it with his own eyes. This short history of a magical stone reveals some of the ways in which the community of seventeenth-century angel conversants operated, and how practical knowledge sometimes traveled among its members.[205] If the vehicles of knowledge transmission among practitioners of astrology were usually written texts (astrological tables, theoretical and practical manuals, letters and manuscripts), there were also active magical objects that could take on this role in the more secretive art of angelomancy.[206]

John Dee possessed a mirror of black obsidian and several crystals that he called "shew-stones," which were placed during his conversations on a wooden cradle, set upon a table of sweet wood.[207] Crystals (especially of beryl) were thought to possess certain occult properties that made them suitable for the task of projecting spiritual apparitions. Often made into the shape of a "crystal ball," the polished stone was sometimes consecrated with prayer or cleansed with incense or myrrh, in an act that somewhat resembles the suffumigation of astral talismans.[208] Ashmole, commenting on Lilly's account of a Suffolk seer named Gladwell, described a beryl "of the largeness of a good big orange, set in silver, with a cross on the top, and another on the handle." Around the beryl were engraved the names of the archangels Raphael, Gabriel, and Uriel.[209] Nevertheless, any reflective surface, including a polished nail or a glass of clean water, could be used for scrying, and, in some cases, the angels even revealed themselves in open-air apparitions.[210] The doctor's own tools for angelic operations remain somewhat obscure. His alleged "call," opening the previous section, suggests that he operated with a crystal ball. An additional clue may be picked out from a letter he received from a person initialed W. P., answering Napier's question regarding his "philosophers berrill glasse." "[M]ake muche of it," the doctor is encouragingly advised, "and treate it as a rare secreat, for it is a most excellent thing and was made for noe private vse of the Artist, but you maye haue onlie the vse of it. it is made for things to come, keepe it as a pretious Jewell, ... [as it] would haue cost you many miles ryding, and muche expence."[211]

Whether they appeared in a crystal, on a reflective surface, or in open air, the question still remains how spiritual beings such as angels could communicate with a human scryer. Often discussed as part of the speculations over angelic corporeality, the ability of angels to produce human-like means of communication was therefore an important aspect of the mechanics of angelic magic.[212] In the face of the many biblical appearances of angels, even those who held that angels are pure spirits still claimed that they could imitate human speech by forming the desired sounds in the air, or imprint their thoughts upon the human mind just like sense impressions.[213] Lilly thought that although angels could mimic a human voice, it was very rare for them to speak articulately, and "when they do speak, it is like the Irish, much in the throat." Instead, he suggested that messages could be conveyed by angels appearing at a distance, "representing by forms, shapes, and creatures, what is demanded."[214]

Attaining Divine Knowledge

Speculation about angels reached its peak in the thirteenth century with the theological debates of Bonaventure and Thomas Aquinas, both of whom had developed systematic, comprehensive angelologies. Their inquiries into the nature of angels helped the understanding of the mechanics of angelic cognition, foreknowledge, communication, and locomotion.[215] Many of these issues were of real theological importance, both in defending the portrayal of angels as ministering spirits and because they directly concerned the economy of salvation (since angels were transmitters of prayers to God). The condemnations of 1277 issued by Bishop Stephen Tempier of Paris, one-seventh of which concerned angels, marked a break in the endeavor to create positive knowledge of the celestial world. Leading figures of the Reformation such as Philip Melanchthon and John Calvin went further in denouncing this "idle scholastic quarreling," no longer relevant for salvation.[216] Nevertheless, Reformed writers continued to investigate angels extensively.[217] Early modern scholars had often very different ideas regarding the nature and epistemology of angels, but in a way revived the same medieval discussions about them as a way to knowing God. A practicing conversant with angels, Napier too attempted to utilize his Jacob's ladder to further probe some of these cosmic obscurities:[218]

> Q: "Wheather any angels were created before this visible world was created." ℞: "some before & some after." And again: "All angels were not created at once, some before the worlds creation, some after."[219]
> Q: "Wheather the places of those angeles which fell shalbe supplyed according to the number of those that stoode & fell not, or according to ther [sic] number that fell." ℞: "according to the number of the[se] that fell."[220]
> Q: "Wheather the good Angels hath 77 names by which on[e] knoweth another, as michael, Gavriel, Vriel [and] Jeremias."[221]
> Q: "Wheather there be more good angels then bad." ℞: "more good."[222]
> Q: "Wheather sathan can be in more places then on[e] at on[e] & the same instant."[223]

One particular Christian tradition that engaged early modern thinkers was the legend of the Antichrist: the creature whose coming was to precede the end of the world. The word "Antichrist" comes from the Epistles of Saint John, where it is applied to those who deny that Jesus is the Savior.[224] In medieval folklore, the Antichrist became a Jew of the tribe of Dan, who would be born in Babylon just before the end of the world, persuade the Jews that he was the Messiah, and rebuild the temple in Jerusalem.[225] In the very early stages of the Reformation, the term was applied to the pope as part of the Protestant campaign against anti-Christendom in Rome, and in post-Reformation England great efforts were still made to prove that the Antichrist could not be a Jew and that he had already come.[226] The years after 1600 witnessed a flood of writing on the topic of Antichrist, and rumors of his sightings in various places became rife, sometimes confused with the medieval myth of the Wandering Jew.[227] The

Interviews reveal Napier's repeated attempts to resolve this ancient mystery through his spiritual conversations with Raphael:

Q: "Of antichrist, who this antichrist shalbe, wheather a turke, pope, or a Jewe" ℞: "a Jewe." Q: "of what tribe." ℞: "of the tribe of dan, a Jew of the tribe of dan."[228]

Q: "When Antichrist shall come, & what shall happen after his death. wheather they Jews shall live in peace & all the gentiles with them. wheather any of the gentiles shalbe also converted."[229]

Q: "What shall happen then & who they be that shalbe converted, wheather only the Jews or the gentiles also that were not before converted."[230]

Q: "How long the world shall last after the death of Antichrist."[231] ℞: "The name of Antichrist is Susei, Jesus [spelled] backward. he shalbe a Jew of the tribe of Dan. Jrama his mother. maria the mother of Jsus christ & Jrama the mother of Antichrist. The name of christ Jesus, the name of Antichrist Susei."[232]

Napier also consulted Raphael on controversial theological issues, especially those that also featured in his correspondence with fellow theologians. For example, he asked the archangel whether souls can enter heaven or hell before the Last Resurrection,[233] and if the bread and wine in the sacrament were actually transubstantiated into Christ's flesh and blood "as the laterane concell affirmeth & so conse[nte]th."[234] As the next chapter will show, this Anglican country cleric did not hesitate to engage in certain religious controversies. Napier never publicly claimed to have received theological insights from the world of angels; yet his spiritual conversations, mostly conducted in the period immediately before he allowed himself to openly advertise his theological stance in letters and tracts, may well have confirmed for him that truth was indeed on his side.

Q: "Wheather any man can be infalliblely sure of his salvation without some speciall revelation." ℞: "he cannot."[235]

Q: "Wheather soules after this life goe to heaven before the last resuration or any to hell before that time."[236]

Q: "Whether Adam had the spirit of God before in the state of Innocency. They say he had not."[237]

Q: "The woman ought to have power on her head because of the angels, in 1 Cor[inthian] 11.10, what is ment thereby."[238]

Q: "Why the old prophet was permitted to deceave gods prophet; Wheather he had a revelat[ion] from an angell or he faynd [feigned] it. Why he was not peanished [punished] for his deceits."[239]

On the more practical side, Napier occasionally applied for the archangel's help in gaining methodological and technical guidelines for both medicine and alchemy. In one such example, he consulted Raphael on whether it would be wise to let blood at nighttime, and if so, whether with a lancet or with horse leeches.[240] He also inquired if bloodletting was appropriate for patients with the purples [purpura] or smallpox, "especially when through there disease they doe likely pour blood."[241] Unfortunately, no answer was recorded for any of these questions. The archangel was apparently more willing to help with questions regarding the use of astrological talismans. "Wheather [a] S[igil with] char[acters] of [sic] on[e] syde be as effect[ual] as on boath sydes," Napier once asked Raphael.[242] "It shall serve on the on[e] syde very well" was the reply. The archangel also supplied some detailed instructions for using a sigil to cure witchcraft: "make agaynst *veneficium* the Rundle of Raph[ael] & stipe it eyther in mug water or *hyp[ericum] cum [precibus?]* & let them ware it 3 days & researve it. vse this as a principall about sigils yet grave it in hora ♃ to make it more Effectuall."[243] As I have argued in the first chapter, horary astrology usually provided Napier the necessary evidence for making a medical judgment. When formulating a cure, however, he sometimes sought additional angelic guidance:

Q: "If after the pox be gone, the party gone lame in there limbes, handes & arms; wheather after a medison to kill the venoume, a purge & a purging dyet drink be not profitable."[244]
Q: "Wheather in knees swelled, what comst best to remedy it."[245]
Q: "In orthnopopeia [orthopnoea] in the [streghtens?] of the lengs [lungs], what best to open it & to mend it."[246]
Q: "a quartyne feav[er], how cured."[247]

The science of alchemy imposed incredible challenges and continuous frustrations on its practitioners, to the extent that the sixteenth-century Italian alchemist Vannoccio Biringuccio came to the conclusion "that unless someone should find some angelic spirit as patron or should operate through his own divinity," the art was destined to fail.[248] Even Meric Casaubon, the Anglican clergyman who publicized Dee's conversations with angels, denouncing him as being misled by evil spirits, was prepared to concede in 1659 the possibility that many chemical secrets "came to the knowledge of man by the Revelation of

Spirits."[249] Deeply engaged in practical alchemy, Napier hoped to make some epistemological shortcuts:

Q: "Wheather the mutch recommended worke of Cako will prove true worke in alkimye [alchemy]." ℞: "It will by gods grace."[250]
Q: "Wheather Mr Cookes *aurum potab[ile]* made after Rhemans will dissolve the whole body of Gold, or wheather his owne worke be better for *aurum pot[abile]*."[251]

Napier's angelic conversations display therefore a plain transmission of essentially practical knowledge, touching those scholarly domains with which he was incessantly engaged: religion, medicine, and alchemy. In all of these, his supposedly direct access to celestial wisdom granted him an additional channel through which certain, reliable information could be gained.

The Fortunes of Angels

To a large extent, Dee's conversations became famous simply because they provided a peek into a uniquely large-scale discourse that was credited to angels. Napier's archive is not only an additional hoard of such documented dialogues, but also the single surviving source for practical angelic divination and its employment in early modern medicine. During his lifetime, however, his endeavours invited the kind of misinterpretation that could lead to the perilous accusation of conjuring. "I received a letter from Mr Spencer . . . of Oxford," Napier wrote in his diary in the spring of 1620, "[who] told me how my lord chancelor extolled me & the Earle of Exeters son called me [a] coniurur & how my lord Wentworth comended me & pleaded for my memory; the lord that knoweth the secrets of all harts reword him & all his family for it."[252] Since, at least by letter of the law, the crime of conjuring evil spirits demanded capital punishment, such allegations posed a real threat for Napier.[253] Although sanctions were in practice quite mild, it might be that they underlay the ultimate disappearance of angels from his medical diaries.

Contemporary assaults on reported angelic contacts were usually fueled by suspicion of "popish" idolatry and disputes over true religion. One major concern regarded those Catholic elements still embedded in the magical ceremony, such as reciting certain formulae of prayer, employing a virgin scryer, or practicing temporary asceticism. While some Protestant clergy viewed angelic apparitions as a way to infuse renewed belief in the face of growing irreligion, more "conservative" churchmen were alarmed by such displays of religious enthusiasm.[254] Others warned that genuine spiritual encounters had long before ceased and that the operations of angels were no longer discernible by human eyes; hence new apparitions almost certainly involved demons in disguise.[255]

Furthermore, invoking angelic assistance seems to have challenged the Protestant principle of immediacy and the superfluity of any intermediaries to God. Not only was the worship of angels seen in Protestants' eyes as robbing God's honor, but even putting trust in these celestial agents was sometimes conceived as derogating Christ's efforts.[256] We might have expected, therefore, a substantial decline in reports of angelic communication in Protestant terrain. Yet, as several historians have recently shown, seventeenth-century England rather saw the blossoming of such alleged contacts.[257] One reason was that, unlike saints, angels could hardly be discarded as nonscriptural, and while their veneration came to be explicitly prohibited, only a few were ready to dismiss the assumption that good and bad angels inhabit our world and could be contacted.[258] In what follows, I shall nonetheless offer a more fundamental solution to this intriguing puzzle.

Thirteenth-century theological debates over angels, it is claimed, arose in part as a response to contemporary discussions of Aristotelian and Neoplatonist cosmologies.[259] Formulated by the sixth-century theologian Pseudo-Dionysius the Areopagite, the Neoplatonic ontology of hierarchical mediation was based upon a view of the heavens as ordered in a concentric pattern. Each wheel in this cosmic model was moved by an individual "intelligence," all emanating from one another and depending upon one principal, self-moving cause.[260] The identification of Christian angels with these celestial "intelligences," endorsed by most medieval luminaries of the thirteenth century, helped establish their role as privileged creatures in the hierarchical structure of the cosmos, superior in nature and wisdom to earthly humans.

The condemnations of 1277 sought to combat this assimilation of pagan ideas into Christian truth. Particularly, it was feared that the identification of angels with the eternal "intelligences" might blur the gap between God and his creation and vitiate his absolute freedom. Later scholars who sought to further explore the angelic realm had to keep these warnings in mind and treat angels on equal footing with other created beings. Anja Hallacker cites three seventeenth-century examples to show how contemporary philosophers and theologians promoted such essential likeness of all worldly creatures. "Arguing in the context of a vitalized universe," she maintains, "... they remove the border between the divine and the terrestrial world by drawing a line of demarcation between an infinite God and the finite creation."[261] Protestant angels were no longer "superior creatures," capable of aiding in human salvation. Yet having retained their elevated cosmological status, they could still mediate divine wisdom to mankind, acting as valuable agents in exploring God's creation.[262]

Napier's appeals to the archangel Raphael suggest that he regarded angels as having privileged access to heavenly inspired knowledge. His spiritual conversations might be fitted therefore within a broader Christian-Neoplatonic outlook, in which divine wisdom emanated by degrees from the first cause downward. In such a cosmic order, archangels, like the planets and stars, were trustworthy informants of divine knowledge. Nevertheless, he too saw his heavenly messengers as standing on equal (if not lower) theological footing with man: "man is more highly esteemed of god, then the angels . . . as first that god soe loved mankind that he gave his only begotten sonne as the price of there redemption, which he did not for the anngels; 2 because he tooke not the nature of anngels and yet would & did take the nature of man in comand & vnited it to his person."[263]

A somewhat related distinction emerges out of W. J. Torrance Kirby's response to the apparent tension in some of the modern assessments of Richard Hooker's thought. Kirby claims that whereas in matters of salvation Hooker was advocating a Christologically centered account of mediation between God and man, in regard to political theology and divine governance he accepted rather the Pseudo-Dionysian hierarchical ontology of divine ordering.[264] Hooker's theology, denying the need for any hierarchical mediation in soteriology, did not exclude therefore the concept of divine hierarchy in ecclesiology and cosmology. Recognizing such dissociation between cosmological and soteriological theories might open the way, I believe, to a better understanding of the persistence of theurgic magic in Protestant lands: one that acknowledges the enduring role of angels in mediating divine wisdom despite their loss of theological privilege.

But why were such techniques employed by Napier in the first place? What purpose did magic in general serve in his medical and divinatory practice? In the preceding chapter, I outlined a preliminary division between Napier's astral and angelic magic, built on the distinction between *acting* and *knowing* in practical medicine. While the doctor's talismanic art allowed him to manipulate the forces of nature, theurgic magic provided him with the means for bridging the gap between flawed human understanding and perfect divine knowledge. Still, both allowed Napier to exploit those layers of the cosmos that were not usually accessible to ordinary humans, enhancing the effectiveness of his practice beyond the typical scope of astrological medicine. The Staffordshire astrologer William Hodges, maintained Lilly, "resolved questions astrologically, . . . [but]

in things of other nature, which required more curiosity, he repaired to the crystal."[265] Napier himself seems likewise to have approached his angels only in select cases—those in which the astrological judgment proved unhelpful or vague. His spiritual interviews thus provided a supplement rather than an alternative to judicial astrology, an extraordinary appeal to a higher instance within the same knowledge frame.

4

RELIGION AND KNOWLEDGE

Alongside his flourishing medical and divinatory practice, our busy minister engaged in routine correspondence with neighboring scholars and ecclesiastics. Some correspondents were former tutors or fellows of Oxford colleges; others became acquaintances after coming to Great Linford to seek medical help.[1] An active figure in local religious circles, Napier had recurrent conflicts with local clergymen and was occasionally involved in oral disputes.[2] Like many other ministers who were not keen on Calvinist ideas, he valued private prayer above the church sermon and, to say the least, was not especially punctilious in performing his preaching duties. According to William Lilly, "Miscarrying one day in the pulpit, he never after used it, but all his life-time kept in his house some excellent scholar or other to officiate for him, with allowance of a good salary."[3] However, since Napier's manuscripts include a scattering of drafts of sermons and contain references to his preaching, Lilly's assertion should be taken with a small grain of salt, or at least restricted to a certain period.

Having little concern for the formality of religious rites, Napier left almost no clues as to how he dealt with most of the period's controversial ceremonies, such as kneeling at communion, wearing the surplice, churching women after childbirth, or marrying with a ring.[4] He did however voice his opinions on sundry doctrinal matters, leaving behind about one and a half dozen letters and short tracts that deal with questions of Christian authority and matters of belief.[5] Some of the letters are part of an ongoing (yet otherwise lost) correspondence, in which one side would answer the previous question and then pose a new one; others are onetime manifestations of Napier's views. For Napier's contemporaries, religion and the investigation of nature were hardly separate spheres, but rather two complementary paths to knowing God.[6] It is therefore my aim in what follows to recognize the theological intricacies and doctrinal premises within which his astrological-medical practice was embedded and justified.

In the first part of the chapter, I discuss Napier's conception of Christian authority and certain knowledge, reconstructing his more or less organized argument about the special authority of the primitive church and of the "general consent" found among its doctors. Like many of his contemporary compatriots, Napier sought to regain the firm ground that was lost when a millennium of Catholic Scholasticism was suddenly rejected. He too found such ground by way of a methodical return to the dawn of Christianity and the primitive church. Yet rather than build his way up from a fresh reading of the Bible and a few early doctors, as some of the more extreme Reformers attempted, Napier embraced the long chain of cumulative knowledge formed by generations of Christian scholars, peeling off only what he considered to be outright "popish errors."

The second part of this chapter explores Napier's engagement in two contemporary theological controversies: the citing of pagan authors in church sermons and the dating of Christ's birth.[7] While the two issues were rather marginal in seventeenth-century English polemics, they provide some useful keys for unlocking his ideas on harvesting true knowledge. When one looks at his theological expositions as a whole, Napier might be counted among those "conservative" English churchmen who saw themselves as distinct from Roman Catholicism, but did not feel obliged to further reform in the church along Calvinist lines.[8] Denying the authority of the pope,[9] he nonetheless used the Latin Vulgate and cited from noncanonical books (though rejecting their canonicity). He asserted the sufficiency of Christ's sacrifice and justification by faith alone,[10] but believed that man was capable of accepting or rejecting God's grace once it was shown to him,[11] that praying for the souls of the just was legitimate if made as an offering of gratitude,[12] and that citing heathen literature in a sermon did not impair the purity of the word of God.[13] And he counted "dressing, kneeling, oyling, fasting, pilgriming [and] praing for the dead"[14] among the errors of papistry, which he condemned as the "church of infidels";[15] yet he sanctioned the use of the cross at baptism and ruled in favor of celebrating Christ's birth date on December 25. His markedly conservative outlook seems therefore to have been reconciled with his genuinely "antipapist" rhetoric, the latter conventionally identified with the "populist" elements in the English church.[16]

Christian Authority

Unlike "popery," which was looked upon as antireligion or as the corruption of the church by the Antichrist, Catholic doctrine, backed up by hundreds of years of human intellectual effort, was a valuable source for the emerging English church. In seeking firm new grounds for religious truth, many English theologians appealed to the authority of the early church fathers and the primitive, undivided church as a counterweight against the authority of Rome on the one side and the individual interpretation of scripture on the other.[17] Exactly how far to reach back was a disputed question. English pro-Calvinists were usually prepared to rely only on scripture and perhaps Augustine. More conservative Anglicans were willing to accept everything that was written by the church fathers up to the fourth or fifth century.[18] In practice, both sides filled their pages with references to the early church fathers and even medieval theologians, just as Scholastic discussion had done for hundreds of years. As I will show, Napier valued the cumulative efforts of outstanding men, especially those achieved within an established "general consent," over individual opinion. Echoing the form of his arguments for the lawfulness of astrology,[19] his view suggests a consistent approach to the attainment of true knowledge in all scholarly domains.

It is generally acknowledged that the English church accepted the authority of the first four (or six) general councils but rejected later Catholic assemblies.[20] Article 21 of the 1563 Thirty-Nine Articles, attentive to the recently concluded Council of Trent, stated the unlawfulness of councils that were gathered without the consent of all sovereign rulers, declaring that they "may err, and sometimes have erred, even in things pertaining unto God."[21] Napier himself set out to challenge the legitimacy of certain Catholic councils in a one-page letter directed to a "much honoured friend,"[22] which proclaimed his opinion of the unlawfulness of three assemblies: the Fourth Lateran (1215), the Florentine (1438–39), and the Tridentine (1545–63). Stressing their political illegitimacy, he also examined their agreement (or lack thereof) with "more ancient authority," such as earlier councils or formerly established traditions.

Napier's letter opens with the Fourth Lateran Council and its decree against the "Albigensian heresy," which stated that the same Creator had first made the spiritual (angels) and corporeal (earth) and then created human beings, who shared both essences of spirit and body.[23] These conclusions, claimed the rector, stand in contradiction to those issued by the Second Council of Nicaea almost five hundred years earlier, which ruled that angels were not entirely bodiless. Napier therefore rejected the legitimacy of the Fourth Lateran as one that "assumeth thence authoritye & power to depose Ch[rist]ian kinges," but also for denying the more ancient resolutions of a certified authority. It is important to note, however, that the Second Nicene Council, the authority of which Napier did not question, was not only outside the first four councils legitimized by the English church, but also notorious in Protestant discourse for authorizing the worship of images of angels and saints.

Similar objections are raised in Napier's assessment of the Councils of Florence and Trent. The Florentine Council, maintained Napier, is "noe lawefull nor generall councell," because it declared the apocryphal books to be canonical and because it abolished several traditional sacraments.[24] The Council of Trent, on the other hand, was rejected for its political unlawfulness: "[T]he Trent councell, to which fewe kings sent any Bishops, Neyther Scotland, England, Irelnd, Denemarke, Germany, Polande, nor the French kinge, but doe contest against, that councell beinge packed by the Titular Archb[ishops], *nomine nonve*, by Spaniards & most Italians little better then that Parasytes [little better than parasites]."[25]

The issue of Christian authority, and consequently of true Christian knowledge, seems to have particularly occupied Napier's attention. In an unusual effort to broadcast his opinions in writing, he added to this short letter on the legitimacy of councils a more sustained discussion in two dedicated yet apparently unrelated texts. The first, a letter sent in June 1621 to Henry Jackson, fellow of Corpus Christi College, Oxford, contended that "controversies [are] to [be] determined by the unanimous consent of the most orthodox auntient fathers."[26] The second, a draft of a three-page letter directed to one "dearly respected" on "The deciding of questions of controversie,"[27] is neither dated nor signed, but is unmistakably in Napier's handwriting and seems to be part of a larger correspondence between the two men. Based mainly on these two texts and on scattered references throughout the rest of his theological writings, I will construct Napier's rather organized argument concerning the special authority of a "general consent" among the early church fathers over the individual interpretation of scripture.

For the Scholastic thinker, the Christian faith drew upon three sources of religious truth: the scriptures, the authority of the Roman church, and, in a more limited way, the private inspiration of a few divine men. Among the most basic instruments in this quest was the employment of reason, a word that summed up the efforts of men to attain knowledge by means of their own cognitive powers. That portion of the Christian doctrine that relied upon reason (or logic) more than it did on actual scriptural proof was called tradition; yet reason could be employed only by the authority of the church, based on its "correct" interpretation of the Holy Bible. The most significant feature of the Protestant Reformation was perhaps Luther's assertion, contrary to this conviction, that the Bible was "its own interpreter" and could be understood by most men when guided by the spirit of Christ. Since the Bible contained all things necessary for salvation, Luther argued, no baptized Christian needed to depend on a higher authority in order to understand and practice his religion. Protestant reformers campaigned fiercely against the application of human reason without sufficient scriptural proof, rejecting large parts of Christian tradition as being nothing but "empty reasoning." The three fountains of truth, they urged, should be replaced by *sola scriptura*: the sole authority of the scripture as it was understood by the individual believer.[28]

The authority of bishops in medieval church councils, and sometimes even that of the early church fathers, which had shaped Christian doctrine for hundreds of years, ceased to be acceptable to many Protestants. Reformers claimed that the Latin version of the Bible, the Vulgate, contained noncanonical books, and as biblical scholarship investigated Greek and Hebrew versions of the Bible and found places of disagreement, scripture itself ceased to be an uncontroversial text to which one could simply point.[29] Within a few decades, Western Christendom, until then run by a kind of a theological Supreme Court, became a sort of ecclesiastical anarchy. Even Luther himself came to realize that some sort of authority must be acknowledged, and eventually drew back from the extremity of his initial argument.[30] Nevertheless, those fundamental questions on the nature of true Christianity that were opened for discussion by the forces of the Reformation could not be so easily repressed.

As England became Protestant, its theologians needed to defend their church against Roman Catholicism by relying on some counterauthority that would, at the same time, support their unique version of Protestantism. Such an alternative source of authority, dealing with all the main topics of controversy, was found in the Anglican Articles of Faith. Some historians of post-Reformation England suggest that Anglicanism invented a unique kind of religious authority: one implied by these collections of articles and used to impose a systematic order on the newly emerging church.[31] The English attitude toward religious truth was, however, quite complicated. A central principle of the English church, the *adiaphora*, maintained that certain doctrines and rituals were not essential to the Christian faith and were thus left to the judgment of the believer.[32] Things that concerned salvation, however, demanded absolute certainty, and Napier too made use of the term "infallibly sure" when engaging in such issues.[33]

Hooker, probably the most significant seventeenth-century English voice on the question of Christian authority, emphasized that scripture contained all things necessary for salvation, but nonetheless presupposed the operation of tradition, human reason, and church authority.[34] Napier too repudiated claims to absolute truth based on the individual interpretation of scripture and argued for its appraisal against some established authority. He opposed what he saw as the Calvinist tendency to read scripture according to "such idle fancies, as have no ground neyther out of gods word, councle fathers, ... nor out of any ground of sence or reason."³⁵ Nevertheless, he rejected the Tridentine assertion that "tradition bee helde equall with the canonicall scriptures."³⁶ Following Hooker and other prominent conservatives, Napier thus censured what he saw as inadequate individual readings of the holy text: "[N]ot only Covell but Sutcliffe, archbishop Whitgift & judicious Hooker, doe in there writings so mutch tax Cartwright, who alledgeth the script[ures] not according to there true sence but rather according to his owne humour."³⁷

In the Roman-Protestant polemic of the sixteenth and seventeenth centuries, "the unanimous consent of the Fathers" (*unanimem consensum Patrum*) was a term applied by the Roman Catholic Church to a situation in which the church fathers spoke in harmony to determine tradition. Although the church fathers did not necessarily reach an agreement on every detail of a specific controversy, and although any one of them was not personally infallible, their general agreement in a religious debate and the resulting consensus in the church were referred to as a "unanimous consent" and were assigned a high value of truth. These concordances among the church fathers were seen as the direct continuance of apostolic teaching, preserved in the church through their rulings and canons.³⁸ Napier's case for the special authority of such consent (as opposed to the individual interpretation of the Bible) is supported by four distinct arguments, presented here with appropriate quotes:

I. Not every biblical exposition is legitimate or acceptable, as some people tend to interpret scripture according to their own personal interests or preset agenda:

    [We are sometimes misled by] a private selfewill which maketh vs to misconster the scriptures many times, as they may best serve our purpose & humour, although sometimes agaynst the true & currant sence of the scripture.³⁹

II. Scripture cannot always be taken literally and might oftentimes confuse even an experienced reader:

    [I]t is a vsuall [practice?] in scripture to speake that of all in generall, which applyeth only to the greatest number, as when St. Paule sayth all seeth the things that are there owne & not the things which are Jesus christs, & yet Paule & the rest of the Apostles are excepted. so in that of genesis 6, Noa is excepted for he was a wise & a perfect man.⁴⁰

[A]s all is not gold that gliterets as gold, soe all is not scripture nor trueth, that hath the shewe & semblance of scripture & trueth. for then heretickes might be as good Christians as these.[41]

III. Judgments based on the opinion of several church fathers, guided by the spirit of God, are always preferable to any individual opinion:

> I see no cause why we should not admyt of there [the church fathers'] finall resolution & determination, when they shall decide there controversies according to the publick spirit of the church, guided & conducted thervnto by gods holy spirit & the light of his sacred words, mutch rather then by the private spirit of any particular man opposing against this publick spirit, iudging *quoad analogiam fidei*.[42]

IV. The unanimous consent of the early church fathers is more than merely human and is thus superior to any other employment of human reason:

> I denye simplie that the vnanimous & most generall consent of the catholick & orthodox church lawfully assembled in gods name ... is to be taken & accepted for mere humane, but ought in sence & reason to persuade a godly sober christian man mutch more affectually, then the private spirit of any particular church.[43]

The authority of the church fathers was at its best, Napier thought, when it reinforced our own scriptural interpretations rather than being the sole reference of any defended position. "We mention ... the fathers in our sermons," he quotes from Archbishop George Abbot's answer to a Catholic challenge, "to show that our expositions of the scriptures are not singular & private interpretations, but sutch as weare receaved in the primative church."[44] Napier acknowledges therefore certain limitations to the reliance on the church fathers' opinions. First, he observes, "it be not in itselfe able to *constituere principium theologica* which properly belongeth to the scriptures";[45] that is, it cannot constitute new principles of religion. Second, as there are points that are not discussed by scripture, so not everything was decided by the early church fathers. Such matters are left to the judgment of the individual scholar, who might prove more qualified than any pontiff: "some lay man not authorised may yet be so excelently learned in the scriptures, that his assertion shalbe more credited then the popes definitive sentence."[46] Napier also rejects any attempt to ground ecclesiastical tradition without sufficient scriptural proof, even if it relied on the church fathers' "general consent": "[N]o vnanimous consent of fathers simply is to be admitted, in case it will presume to iudge without or besides the scriptures, for in sutch a case I should prefer the sentenc of a truly a [sic] private man alledging scripture for his praye before a torrent of doctors not regarding or declining from the scriptures."[47]

The significance of the questions discussed here extends far beyond the immediate controversy over the correct way to interpret scripture. Projected onto every theological debate by setting the legitimate scope of discussion, they served as a baseline for most contemporary religious and political disputes. "It would seeme to him that with an indifferent eye should scanne your letters," Napier rightly observed in his letter to Jackson, "that you regard not so mutch to knowe, which abundantly you doe, what eyther protestant or papist holde, as to see vppon what ground or reason they maintayne this their assertion."[48] Indeed, he too was deeply absorbed in questions of Christian authority that defined the way to achieve certainty of knowledge and gave meaning to all religious controversies.

Matters of Faith

Pagan Words in the House of God

"They that will not receave thy gloses vppon the scriptures as the gospell of Christ," Napier protests in a surprisingly long and marvelously antagonistic treatise, "are leekly [likely] to prove in thy iudgment very reprobates & damned creatures."[49] The rector's theological expositions, even when polemical in nature, are normally delivered in a temperate tone that is highly atypical of seventeenth-century religious debate. All but this one particular text. Titled by William Black, in his catalogue of the Ashmolean manuscripts, "An argumentative and apologetical discourse, written by Richard Napier, M.A. Rector of Linford, against the minister of Newport,"[50] the tract is neither dated nor signed. Its attribution to Napier recurs in Frederick Bull's *A History of Newport Pagnell* (published 1900), although Bull admits that the text does not contain anything tying it to the doctor. Bull also dates the tract to the early 1630s, but gives no grounds for this additional inference.[51] What evidence do we have, then, to show that it is indeed the work of Napier? First, the tract's handwriting is clearly that of Gerence James, Napier's most prominent assistant.[52] Second, the only two geographical places mentioned in relation to the events described in the tract are the small towns of Olney and Newport-Pagnell, both in the immediate area of Great Linford. Third, the text is intertwined with numerous references to the same church fathers and theologians that Napier employs elsewhere in his theological writings. Lastly, the style and wording used by the writer, by his own admission "one of the meanest of God's ministers," are very similar to those in Napier's signed manuscripts.[53] Overall, I believe, these points provide sufficient reason to accept the attribution of the text to Napier.[54]

So what was it that upset Napier so much and motivated him to write such an extensive, forceful discourse? From the bits and pieces of information scattered throughout the text, it appears that the unfortunate affair began when Napier was invited to give a sermon in nearby Olney, in the presence of fellow ministers from adjacent parishes. In the course of his sermon, Napier named some pagan gods and goddesses in order to illustrate the ungratefulness of the gentiles toward God: "[H]ere indeede I alledged that the wise men of the gentiles ascribed there benefitts to there fayned gods, as health to there god Jupiter, wisedome to there god Apollo, eloquenc to there god Mercurye, to Minerva learning, to Juno ritches [and] to there goddesse Venus beutye. [A]ll which I proved to be vnthanckfull, because very unthanckfully they ascribed that which was due only to the immortall god, to ther idoll & fayned gods."[55] At this point in the sermon, or very soon after, one of the attending churchmen stood up, accused Napier of heresy in fiery language, and then with some show walked out of the church.[56] Later in the course of the sermon, expounding upon Psalms 119:9 ("Wherewithal shall a young man cleanse his way?"), Napier cited an appropriate passage from Horace, which triggered further fury.[57]

Later that week, at a dinner attended by some of the area's most eminent residents, Napier sat silent, disgraced and humiliated, as he was crudely condemned for his sermon in front of the other guests. Yet this was hardly the end of the ordeal. The minister who had furiously left the church during Napier's sermon issued a written attack, in which he accused the old rector of unlawfully alluding to fictitious, profane stories in the house of God.[58] Perhaps worried that the subject of his righteous wrath would decide to fight back, he sent one of his acquaintances to Napier in an attempt to persuade the doctor to acknowledge his errors and avoid further confrontation. Yet such a mortifying attack could not be left unanswered, or so Napier felt: "as if a waspe should crave he might stinge some naked bodye & the partye should sit still & say or doe nothing."[59] Determined to clear his name and correct the perceived injustice, he composed an open, learned answer to the charges leveled at him.

Ann Hughes, investigating the tense religious atmosphere in the slightly later period of the interregnum, has shown how rival accounts of the same debate attacked the truthfulness of their opponent's version.[60] Napier, no exception to this rule, accused his rival of misinterpretation, distortion of the truth, and downright lies. Nevertheless, this written exchange constituted a solemn dispute between two learned ministers. Napier's apology, in particular, is organized according to the same *quaestio* form developed over centuries of Scholastic polemics and still used in seventeenth-century England in formal written disputations.

The image of Napier's adversary, as it emerges from the text, is that of a typical "Puritan":[61] zealous and quick-tempered, anxious to express his fiery fervor and intolerant to the opinion of others. The most salient feature of people labeled as "Puritans," it has been argued, was that they appeared to their less zealous brethren as "self-righteous and self-selecting groups whose high opinion of themselves made them the object of a good deal of ill feeling and satirical humor."[62] More conservative theologians often felt that their opponents were prepared to openly defame them in order to enhance their own reputation for zeal and godliness.[63] Napier seems to express the same feelings when he announces that the intention of his tract is "to cause the zealous & hot spirited brother ever after to take heed howe his zeale burne, least vppon better vewe [view], it seeme to be but *ignis fatuus* [an illusion, lit. foolish fire]. [Y]et because they that are wedded to faulse & selfe love, are soe blinded in iudgment that they cannot or willnot discerne truethe from falsehood, but embrace every shewe & semblance of trueth for the substance thereof, as Ixion insteed of Juno embraced a very cloude."[64]

Pro-Calvinist clergy—perhaps the more appropriate characterization of Napier's adversaries—often quoted the Bible as uncompromisingly against myths and poetical fictions, emphasizing the danger in their allegorical rendering. Their campaign was mainly based on several short passages from 1 and 2 Timothy, which were also used as the foundation of the attack on Napier.[65] However, alluding to heathen myths in a church sermon was not an uncommon practice. Gervase Babington, who was successively the bishop of Llandaff, Exeter, and Worcester, quoted extensively from an ancient tale on the ungrateful Nephastes in a sermon he gave in May 1591. "[B]e it [a] storie or [a] fable," he explained, "the drift of it is Gods truth, as true as God, that unthankfulnes is obvious to the Lord."[66] Such references to heathen literature rested upon a long-established tradition that originated in early Christianity and was further developed with the Renaissance reawakening of Neoplatonic ideas.

Elements of "pagan wisdom," including philosophical thought, natural philosophy, arts, and mythology, were interwoven into Christian learning almost from its very start and challenged the new religion's claim as the sole source of knowledge in all aspects of life.[67] Greek and Roman mythology posed a particularly difficult challenge for Christian educators, since its literal meaning was extremely problematic from a moral perspective. Still, interpreting mythology in a way that endows it with educational meaning goes back to the pre-Christian Stoics, who tried not only to portray the gods as mere symbols of the powers of nature but also to uncover the spiritual significance concealed in their allegedly immoral adventures.[68] Both allegory and parables were standard techniques in early Christian teaching, and the patristic tendency to look for hidden meanings behind scriptural narratives made it hard to condemn the same method when applied to myths.[69] As a means of education, it was realized that, like the Bible, pagan myths and poetry used the cover of fables and imagery to pass on the message of truth in a more suggestive way. Commentators and preachers in the Renaissance alluded to classical poetry and prose when reminded of a classical reference, or simply felt that what they wished to say had been better phrased by a classical author.[70]

Christian scholars, however, discovered in mythology much more than allegorical interpretations; they found within it the Christian doctrine itself. The feeling that classical myths were not all fiction goes back to the first centuries of Christianity and persisted through generations of learned scholarship.[71] Ancient myths came to be viewed as dim or distorted memories of either confused or intentionally hazy truths, promoting the idea that under the varied forms of all religions was a common kernel of truth.[72] Certain characters and episodes of pagan literature were found to prefigure Christian narratives, in the same way that the church fathers had pointed to characters and episodes of the Old Testament as heralds of Christ's life events. This doctrine of parallel revelation would later strengthen the Neoplatonist comprehension of the universe as a great myth, within which the Creator had scattered numerous hints in various places. Prominent Renaissance figures such as Marsilio Ficino and Pico della Mirandola promoted the belief that ancient poets, philosophers, and magi were divinely inspired and, like biblical prophets and apostles, had certain access to heavenly knowledge. Pagan mythology became thus not only a source from which distorted truths could be "dug out," but a sacred text endowed with spiritual meaning and religious axioms, which invited the employment of scholarly exegesis.[73]

In his fondness for allegory and concealed truths and his inability to forgo an opportunity to quote a few lines from Horace or allude to an appropriate fiction, Napier fits well into this tradition of exegesis through pagan literature.[74] His broad humanistic education and his acquaintance with the literature of Rome and Greece are very much apparent, and he definitely saw a moderate use of pagan testimonies by way of comment on the scriptures as not only lawful but commendable. Time and again, he praises allegory and its value over literal reading, suggesting for the former a direct scriptural foundation: "It be recorded in Nehemy 8:8 howe that they [sic] levits reade in the booke of the lawe of god distinctly & gave the sence therof & caused them [the people] to vnderstand the reading."[75] Heathen literature, Napier argues, is not just a substantial part of an appropriate humanist education, but an essential tool in the hands of the educated minister: "[W]e gather how well it agreeth with the minister of god that he be armed & stored not only with the testim[onies] of the script[ures], but also with the testim[onies] of the heathens, that he may applye him selfe to them, when he ought to amende & to edifye vnto godlines."[76]

Napier thought that, as a tool of religious teaching, pagan literature could and must be employed to achieve educational goals. "Augustine willeth & counselleth vs," he reminds his adversary, "that we shall rob the heathen of ther ritches and translate it to the garnishing of the church."[77] There is no reason therefore why the marrow of heathen myths should not be exploited for ornamenting a church sermon. Intended first and foremost to illuminate scripture, they might bear an important message masked as fantastic fiction. The gospel itself has shown how the events of Christ's life appear in the Old Testament. In a similar way, Napier explains, he had repeated a familiar rendition of Hercules's struggle with the Hydra as a means of glorifying Christ: "I have nothing els but sought howe I might rightly & truely apply, that which others have misapplied to our victorious Hercules, to that lion of the tribe Juda[h], that is to our christ, who hath not in a fiction . . . but in very trueth he [sic] subdued sathan."[78]

Napier also shared the belief in the divine origin of heathen testimonies. "[B]e it story or be it a fable," he declares, "the drift of it is gods trueth"; and again, "our fictions, never so fabulous, yet the drift of them is as true as the script[ures]."[79] He quotes Tertullian, who "sheweth that they heathen have many true things in them which either them selfes or ther ancestors had out of the scriptures."[80] Napier saw in the classics a source of divine wisdom that might be inferior to the scriptures but was nonetheless valuable and instructive. Fictitious fables, properly allegorized and explicated, may thus convey God's truth: "for poeticall fictions & fables with there fitte & orderly applications, always agreeable vnto the worde, may welbe seene the ambassadour of god."[81]

Rather than relying on the monopolistic voices of the scriptures and the church, Napier saw a multitude of divine channels that together composed a spiritual whole. In this he seems to have followed the Neoplatonic tradition that looked upon a world filled in every corner with elements of divine revelation. Using a particularly beautiful image, he explains how these disparate channels relate to one another, "strongly wouen as it were, on[e] thred within another, that a man can hardly heave [have] the voice of the spirit without his Echo, one place leeke [like] the mathematicall paralels, equally correspondent, running in a circle, all meeting togeather in the same centre, expounding on[e] another, explaining on[e] another, joyntly affirming the same trueth."[82] Since the word of God is in itself perfect, it is not complemented by these other means of revelation, only expounded and illuminated. "[T]o adde beuty to the word,"[83] Napier declares, does not impair its sufficiency and purity: "[b]oth heathen testimonyes & poeticall fictions, soberly vsed & fittly applied, may well stande with the worde, notwithstanding the purity, sufficiency, or any other title of commendation due to the worde."[84] Allusion to heathen testimonies does not imply, he repeatedly stresses, their mixing with the pure word of God or the "fearfull coupling of Baal-Peor with the mighty Jehova."[85]

Prominent Reformation leaders, including Martin Luther, John Calvin, and Philip Melanchthon, denied the legitimacy of allegory and insisted on the primacy of a literal reading of scripture.[86] The sixteenth and seventeenth centuries thus saw the gradual stripping away of symbolic elements and the demise of allegorical interpretations.[87] This affected enormously the incorporation of pagan literature into Christian discourse, as well as the discussion of how to read nature.[88] Napier represents therefore a tradition already on the wane; yet for him, tropological and allegorical readings of the book of nature were just as admissible as such readings of the book of scripture. Allegory played a vital role in his conception of the universe, both in reading the picture of the heavens and in manipulating the power of heavenly bodies into terrestrial objects (i.e., sigils). Its rejection—essentially a denial of the ability of things to act as signs—had obvious consequences for the role of the stars in practices such as his.

Dating Christ's Nativity

"[Y]ou have taken extraordinary paines in answeringe my last demands," wrote Napier in January 1629 to some "worthy Sir," "touchinge som[e] seeminge differences about Peters Universality, whether it so agree to him that [not] onely the laity taught by him, but also [those taught by] the rest of the apostles his fellowes, should bee called *oves s[anc]ti petri, cui cura incumberet pascendi illos uti oves hujus pastoris.*"[89] Satisfied with the answer he had received, the rector moved on to answer the question posed back to him, regarding the exact date of Christ's birth. This query is the main theme of Napier's letter, and although it is unclear whether he had any additional reasons for engaging with the subject, its treatment can shed further light on his theological and philosophical outlook and on his ideas on the means of true revelation.

The celebration of the Savior's birth, lacking any explicit scriptural reference, was established in the Western church sometime around the fourth century, following a series of expositions by the church fathers. Some of these church fathers, including Augustine, Chrysostom, Gregory of Nyssa, and Ambrose, concluded in favor of December 25, probably reconfirming an already existing tradition. Others, such as Clement of Alexandria, Epiphanius of Salamis, and Paulus Orosius, insisted that it fell on other dates. To ground his conclusion, Augustine argued that John the Baptist was born at the season of the year when the days start to shorten, namely, about the summer solstice in June. Christ, conceived six months later, was thus born when the days start to lengthen, around the winter solstice in the later part of December.[90] Chrysostom, taking a different route to reach the same conclusion, placed John's conception in September, when his father supposedly completed his duties as the high priest in the temple. A third line of argument involved a tortuous computation starting at Christ's baptism, believed to have taken place exactly on his thirtieth birthday, and calculating from thence the exact duration of each of the following events of his life to the Passover in which he was crucified, amounting to three years and three months. Going back three months from Passover, one arrives again at the end of December.[91] The voices of opposition were apparently overcome, and echoes of this early dispute had largely subsided by the Middle Ages, making the celebration of Christmas on December 25 a firmly established tradition in Western Christendom.

Protestant reformers, seeking to abolish any Christian tradition suspected of having sprung from pagan sources, attacked Christmas on dual grounds, arguing that there was no scriptural directive to celebrate the Savior's birth and that its dating to December 25 was completely unfounded. They also pointed out that some ancient pagan religions celebrated the rebirth of the Sun on the same date.[92] The chief argument employed against observing Christ's birth on December 25, or at any other date for that matter, was the lack of any explicit scriptural basis. Some scholars, such as Joseph Scaliger, Henricus Wolphius, Sethus Calvisius, John Lightfoot, and William Drake, suggested alternative dates in September, October, April, or May.[93] Others did not offer a specific alternative, but rather presented a wide range of possibilities. Their aim was not to set another celebration day in the place of Christmas, but to abolish this erroneous tradition altogether.[94]

Napier's letter begins with the multitude of alternative dates suggested by the church fathers and more recent theologians: January 5 (Epiphanius), May 14 or April 3 or 4 (Clement of Alexandria), April 8 (Paul of Middleburg), or sometime in the autumn (Philippus Beroaldus). Yet "the Westerne Churches," he stresses, "doe most accord in this that Christ was borne as on about the 25 of December, which is most usually observed in all the Westerne parts of the world."[95] Napier repeats Henricus Wolphius's qualitative reasons for rejecting Christ's nativity in December: it was no fit time to travel in the midst of winter to Jerusalem in order to pay tributes, and no shepherds or sheep could endure such cold weather.[96] Nevertheless, he argues, "thes[e] reasons of his weigh too light to condemne the learned doctors of the western Church who scorne to yeild to such reasons, though somewhat probable."[97] He also mentions Augustine's conclusion that Christ was born in the time of year when the days begin to lengthen, but admits that there was some disagreement about this among the church fathers. Once again, Napier's distaste for individual objections to generally consentient ideas is revealed, and it is very much consistent with his ideas on how to evaluate religious truth that we examined in the first part of this chapter.

There was also an entirely different kind of reasoning that could be exploited in this specific dispute; one that involved the deciphering of natural hints rather than the Scholastic maneuvers of biblically based arguments. A famous passage in Numbers declares, "There shall come a starre out of Iacob, and a Scepter shall rise out of Israel, and shall smite the corners of Moab, and destroy all the children of Sheth."[98] This passage, often quoted by Christian writers as prophetically referring to Christ's birth, also appears to suggest the concurrent picture of heaven. Another indication of a special heavenly constellation at Christ's nativity appears in a passage from Matthew describing the arrival of the magi immediately after the Savior's birth: "for we haue seene his Starre in the East, and are come to worship him."[99] Christ's nativity was shown by this constellation, maintained the author of the *Speculum Astronomiae* (traditionally identified as Albertus Magnus), not because he, who created the stars himself, was subject to their motion or judgment, but because the book of nature could not be complete if it ignored its own Creator.[100]

The famous ninth-century Persian astrologer Abū Ma'shar al-Balkhi (Latinized as Albumasar), following by his own testimony the teachings of the "ancient Chaldeans," described a virgin and her son, "whom many people call Jesus." This "virgin" was associated with a particular constellation that rose with the first "face" (the first ten degrees) of the zodiacal sign of Virgo. Based on this text and on the pseudo-Ovidian poem *De vetula*, the philosopher and Franciscan friar Roger Bacon (b. 1214) concluded that Christ's birth had been foretold by the stars.[101] The *Speculum Astronomiae*, discussing Albumasar's views on the nature of planetary motions, consequently ruled that Christ was born when Virgo was in the eighth degree of the ascendant. The Italian physician and astrologer Pietro d'Abano, his countryman Cecco d'Ascoli, and the French cardinal Pierre d'Ailly were among the prominent medieval figures who followed this engagement with the horoscope of Christ. Their efforts, and those of others who were prompted by their work, focused on the chronological task of determining the Savior's year of birth and on establishing its calendric date.¹⁰² At the turn of the sixteenth century, Christ's geniture was still a contested ground in which the most advanced scientific tools of astrology and chronology were utilized.¹⁰³

For Napier, a student of astrology and divination, such a route of investigation could hardly be disregarded. "The most learned mathematicians that ever haue been," he maintained, "will haue it to light upon the first degree of ♑, eetching well forward to the second degree."¹⁰⁴ Among such scholars he counts Albumasar, Albertus Magnus, Pierre d'Ailly, and the chronologer and astrologer Heinrich Bünting. Napier's referral to astrological reasoning was not atypical of contemporary discussions of Christ's nativity. Replying to "two over-confident pamphlets," John Collinges, an English theologian and keen controversialist of the second half of the seventeenth century, who took astrology to be more reliable in such matters than any human tradition, complained likewise: "I expected a proof by Astrologicall demonstration, for to this proof by mens saying that lived some 300, some 400, some 1000, some 1400, years after, it is [an] answer good enough to say [that] others [just] as ancient thought otherwise."¹⁰⁵ Others, nonetheless, saw the integration of arguments from natural philosophy into the realm of religion as plain heresy: "As for astrologers . . . who by their horoscopes and calculations have undertaken to declare the day and hour of Christs birth," declared the Massachusetts minister Increase Mather, "their attempt is justly charged with not only vanity but impiety."¹⁰⁶

The employment of astrology in order to extend the scope of human knowledge and solve particular theological issues was hardly limited to Christ's birth. In his "Defence of Astrology," Napier claimed that "the right observation of feastes, especially of Easter, . . . could not be duely kept without the knowledge hereof [i.e., of astrology]," and that astrology "is so necessary to divinity that the booke of Jo[b] without the knowledge of this science could not be well understood, much less rightly expounded."¹⁰⁷ The invocation of astrology in theological debates drew upon the conviction that the stars and other physical objects were tokens of spiritual truths and that, like heathen mythology, they acted as a source of divine revelation that could be read and interpreted in much the same way as the Bible.¹⁰⁸ Early modern chronologists such as Scaliger attempted thus to date biblical events, including the date of creation, the flood, and the Tower of Babel, using astronomical phenomena. Scaliger even argued that the Bible was neither complete nor self-contained as a history of man and should be backed up with evidence from external texts.¹⁰⁹ Laura Smoller has shown how, in late medieval and Renaissance occult thought, scripture, extrascriptural prophecies, and astrology were all used for the same purpose of understanding the secrets of God's creation.¹¹⁰

Napier did not share Scaliger's views on the incompleteness of scripture, but he strongly believed that God had implanted extrascriptural knowledge at creation and had revealed the way to read these messages to select human beings. Moreover, the fact that the word of God was perfect did not imply the same of its human interpreters. Although scripture contained the complete divine scheme, astrology, chronology, and natural philosophy could sometimes prove better tools of exegesis. This rather broad view of Christian truth is stressed in the letter's concluding statement: *"unicuique in sua arte (vt inquit Aristoteles) credendum."*[111] At least in the art of dating the past, Napier seems to suggest, the judgment of the astrologer might be superior to that of the biblical exegete, both being merely human experts in their domains.

In an age still embracing the compatibility of natural philosophy with the Christian faith, Napier was able to substantiate his reading of scripture through natural modes of revelation and, at the same time, legitimize his practical endeavors by appealing to the biblical text. This allowed him to acknowledge astrology as leading to realizations of a theological nature, as well as to justify its use on theological and scriptural grounds. Such a versatile approach to the scope of true knowledge and the application of different texts, signs, and messages as its sources allowed Napier to exploit a broad range of reliable information when engaging in both religious controversy and medical work. Acting as an enabling rather than a restricting force on his daily practice, his theology informed his natural philosophy, and vice versa.[112] Together with a true spirit of active inquiry, it guided Napier's resourceful approach to the exploration of both nature and God.

CONCLUSION

This study has been a journey into the lifelong venture of a single man via the manifold aspects of his practical efforts and intellectual outlook. Throughout, I have delved into an investigation of medicine, religion, and magic, attempting to break, at least to a certain extent, the silence surrounding early modern practice and its practitioners. Former historical accounts of Napier and his like argued for an outright conflict between the practice of "old magic" and some of the fundamental premises of Protestantism, which emphasized more open, direct ways of communicating with God and the universe. This raised further questions as to the ability of some early moderns to reconcile competing intellectual paradigms running contemporaneously during their lifetimes.

A devout Anglican vicar and a popular astrologer-physician and magus, Napier seems a fascinating case of early modern eclecticism. Was this the result of a purely practical agenda, by a practitioner who was intentionally indifferent to, or intellectually incapable of grasping, the underlying theoretical frameworks of his art? This book refutes such a claim. Although not a typical "learned physician," Napier was not an uneducated practitioner and certainly not a quack. University-educated and a diligent reader of numerous ancient and modern texts in theology, medicine, alchemy, and natural philosophy, he was a learned scholar who took part in contemporary disputes and discussions. While his motivations were mostly practical, Napier was certainly concerned with the rational theorization of his practice and well acquainted with the intellectual quandaries of his time. He was perhaps what Eric Ash has termed a "technical expert": pragmatic and action-oriented, yet distinguished from the mass of common healers and popular diviners by his comprehension of how things worked.[1]

How then did this country rector and medical practitioner bridge his seemingly contradictory outlooks and the inevitable gap between his ecclesiastical role and some of his evidently arcane practices? My major conclusion is that Napier not only exhibited practical rather than theoretical eclecticism, but actually incorporated Scholastic, Neoplatonic, and Reformed ideas into a consistent view of religion and nature. Skillful reading in the book of nature, conversing with angels, and scriptural and extrascriptural exegesis all fused into one Christian narrative that inspired the use of varied techniques. Primarily investing himself in the practice of medicine, Napier affords therefore a rare glimpse of a person whose actions, much more than his philosophical or theological expositions, document a coherent worldview.

Adding to something that is already perfect, declared Napier in his defense of citing heathen testimonies in the house of God, impairs neither its sufficiency nor its splendor. Even though scripture in itself is faultless and encompasses all aspects of creation, he believed that God had implanted extra elements of divine wisdom both on and above the earth. These divine messages were placed in the hands of select adepts and heavenly angels, in the courses of stars and the alterations of weather, and in the properties of natural objects. Deciphering profane wisdom, as well as practicing such arts as astrology, alchemy, and angelomancy, was not essential in the quest toward certain knowledge, but sometimes offered a simpler, faster, and perhaps more efficient path to this goal. Scripture itself had acted as the chief instrument of divine revelation, but, as Augustine had already argued, God also spoke the language of things. This early Christian heritage was further developed by the Neoplatonic conception according to which nature, by employing material things as symbols of universal truths, emanates the experience and knowledge of God.[2]

Emanation, in the sense of communicating heavenly wisdom and gaining knowledge of God, was a reciprocal form of communication that allowed humans to participate in the divine. This cosmo-theological concept is perhaps the best string to pull to get at the whole of Napier's theoretical fabric. Fueled by the renewed interest in ancient ideas, it underlay his adherence to the truth invested in various traditions and arts. It also inspired his use of theurgy and contemplation as means of ascending the stairs of creation in order to converse with angels, whom he thought had better access to celestial lore. Emanation proclaimed the active participation of the deity in every aspect of creation and, at the same time, provided a glimpse into the divine order and the nature of God.

Rather than discarding the wisdom of the preceding millennia as built on total perversion and error, Napier adjoined traditional modes of rational thought formed by Scholastic learning to a complex synthesis of orthodox ideas, a Reformed outlook, and Christian-Neoplatonic views. He believed that the most certain truths were grounded in the "unanimous consent" of persons and generations, whether in religion or in natural philosophy. Individual opinion, by contrast, was generally suspect, even if substantiated by additional proof. Like Sir Francis Bacon, the lord chancellor who once extolled his work,[3] Napier was concerned about the new tendency to decentralize knowledge, which was guided by Protestant reform and its emphasis on the ability of individuals to reach truth by their own efforts. His inherent awe in the face of certified, consented knowledge fused, however, with a true spirit of open inquiry and a conviction as to the worth of personal experience.

Committed chiefly to the search for true, action-oriented knowledge, Napier experimented with courses of medical treatment, with novel alchemical and mineral drugs, and with interviews with different archangels.[4] Attention to follow-up and outcome in medical treatment, references to former visits, retrospective notes about a patient's fortune next to his medical notes or angelic prophecy, and scattered recordings of clients' feedback allowed the constant refining of his medical practice and the establishment of acknowledged expertise. A diligent observer, he sanctioned the evaluation of textual authority and ancient opinion in light of the "daily experience" of able men such as himself, promoting an empirical occultism that nonetheless drew on a long-established literary tradition. The outstandingly scrupulous documentation of methods and results within a broad, varied practice successfully broadcasts the practical usefulness of his daily efforts, which offered a viable alternative to the later triumphant modern science.

Napier's burning desire for certain, true knowledge and for acknowledged medical success directed much of his life's work and resulted in an enviable toolbox of practical expertise, technical skills, and know-how. Rather than seeking a name for himself as an ingenious scholar, he chose to advance a purely practical agenda, utilizing existing knowledge to meet the needs of his expanding clientele, yet insisting on its appraisal in light of direct personal experience. In an atmosphere that encouraged creative practicality and a striving for progressive achievements, this made him a renowned physician and a true natural philosopher. Armed with substantial knowledge in theology and natural philosophy but with a heart for the practical arts, he shared some of the goals and priorities of those precursors of "new science" who sought to regain certain knowledge of nature and promote its practical use. Yet unlike some of these men, he favored the vitalist view of an animated cosmos over a mechanistic-materialist model of nature,[5] and a reliance on the cumulative intellectual endeavor of ancients and moderns over the complete philosophical destruction of old learning.[6]

Throughout this book, I have made repeated references to Napier's correspondence with fellow practitioners and scholars, providing a sense of his engagement in various "communities of experts."[7] These overlapping circles of social and professional interaction, whose members shared philosophical attitudes as well as practical interests, document an early modern exchange of ideas regarding religion and nature. The numerous letters about astrology and medicine, the means by which Napier and his colleagues consulted each other on theoretical and practical intricacies, demonstrate how seventeenth-century knowledge circulated among men of practice and how techniques and applications were transmitted from one practitioner to another. His correspondence with esteemed theologians, who urged one another to confront controversial themes of religion, reveals that the scope of English theological discourse ranges far beyond the boundaries of the universities and high clergy.

The breakdown of consensus over Aristotelian cosmology, it is nowadays argued, was the precursor of change in early modern views about the natural world.[8] Yet the collapse of existing philosophical structures did not immediately result in the crowning of a new paradigm or a single methodology. While traditional historiography tended to emphasize the incessant tensions between pioneering mechanists and old-fashioned Aristotelians, between reformed, learned medicine and traditional popular healing, and between a scholarly elite and a superstitious magic-prone populace—all setting the boundaries of man's purposeful journey toward modern science—recent scholarship has increasingly drifted away from such grand narratives. Although emphasizing the already recognized heterogeneity of the seventeenth century's theoretical and practical landscape, my examination of Napier shows how an animist view of nature, an inherent awe for authoritative knowledge, and ideas that grew out of extensive daily experience came together to form a consistent view that injected fuel into a flourishing seventeenth-century practice of astrological medicine. It also portrays this era's unique philosophical outlooks, in essence: a distinct synthesis of classical and Renaissance ideas adapted to a new intellectual atmosphere and its vital currents of natural exploration.

The blurring of historical categories that were once distinct, such as science, magic, or Anglicanism, need not imply that Napier in the seventeenth century saw fragmentation all around him. Attentive to actors' categories, I have shown how his amalgam of medicine, religion, and magic constituted a perfectly coherent worldview for him and for his contemporaries, and how religious conviction did not inhibit but actually stimulated scientific activity.[9] We have here therefore an argument for continuity rather than change and also for the continuously fruitful relationship between religion and science. Overtaken later by the greater intellectual and social forces of a disenchanted world, the doctor's legacy was thoroughly tarnished, and his efforts lumped together with the other credulities of his era. Yet the popularity of his expertise and his participation in busy virtual communities suggest that Napier's cosmo-theological outlook, as well as the techniques and methods that he utilized, was neither extraordinary nor outdated, but simply a coherent strand of activity and discourse in early modern culture.

NOTES

INTRODUCTION

1. Elizabeth was the daughter of Sir John Jennings (or Jenyns), recorded as a lunatic, and his wife, Lady Dorothy—widow of John Latch and the daughter of Thomas Bulbeck, a courtier to Elizabeth I. Burke, *History of the Commoners*, 3:583. The chief source for this case is Elias Ashmole's comprehensive report in London, British Library, MS Additional (hereafter Add.) 36674, fols. 134–37. See also L'Estrange Ewen, *Witchcraft and Demonianism*, 239–40; Linton, *Witch Stories*, 227–28; Wright, *Narratives of Sorcery*, 2:296–98.

2. The literature on witchcraft and accusations of witchcraft is enormous. A seminal book on witchcraft is Macfarlane, *Witchcraft*. On the subject of witchcraft accusations in a medical context or in a particular setting, see, for example, Bonzol, "Medical Diagnosis"; Gaskill, "Witchcraft in Early Modern Kent"; Sawyer, "Strangely Handled"; Sharpe, *Witchcraft in Seventeenth-Century Yorkshire*.

3. This was, most likely, Doctor Simeon Fox (or Foxe), a fellow (and later president) of the College of Physicians of London.

4. Margaret Russell was apparently an acquaintance of the Jennings family and took part in treating Elizabeth. She was nicknamed "Countess" after her namesake, Margaret Russell, the famous Countess of Cumberland.

5. Add. 36674, fol. 134v. "All th'other" are presumably some of Elizabeth's siblings who had already died as a result of this alleged witchery.

6. Napier's medical notes for this case, apparently the preliminary source for Ashmole's account (together with extractions from the formal court records), are in Oxford, Bodleian Library, MS Ashmole (hereafter Ashm.) 222, fols. 15v–16v. A note from April 25, the very day on which Elizabeth accused the four women, seems to have escaped Ashmole. In this earlier consultation, Napier inspected the girl's urine, sent to him by Lady Jennings, but made no mention of suspected witchery. Ashm. 223, fol. 70v.

7. Ashm. 223, fol. 70v.

8. Ashm. 222, 15v. *Epilepttia matricis* (epilepsy of the womb) refers to a disease more commonly known as "the suffocation of the mother" (or simply "the mother"). "This terrible sicknesse is in many things most like to the falling euill Epilepsia," wrote the sixteenth-century German physician Christopher Wirtzung in his compendium of practical medicine, and "may be caused through the retention and putrifaction of the seed, all manner of troublesome accidents, as giddines and paines of the head, madnes, short breath, and panting of the hart." Wirtzung, *General Practise*, 489.

9. See chapter 3 for a comprehensive discussion of Napier's conversations with angels and their role in establishing his reputation.

10. Ashm. 222, fols. 16v, 195.

11. Ashm. 222, fol. 196. The latter remark is in a different ink. The angel's full judgment reads, "halfe a yere henc moriet[u]r venef[icium] patit[u]r sed non a Countis sed ab alijs quam. post illius mortem bene vivent sine omni malef[icio]" (Will die half a year hence. Suffering

[from] witchcraft, not by Countis but by others. After her death they will live well without witchcraft). I am grateful to Dr. Robert Ralley for his help in deciphering this particular text.

12. On the basis of Ashmole's laconic testimony that "the Dr. calls the disease *Epileptica Matricis*, and *Morbus Matricis*," most secondary sources proclaim that Napier dispelled all suspicions of witchery and treated Elizabeth with conventional means. As I have shown, the facts of the case were somewhat more complicated.

13. Ashm. 212, pp. 108, 140–41 (quote at 141).

14. The simultaneous reference to allegedly competing medical theories was certainly a habit of popular healers in the early modern period, but was also useful for learned physicians, who often relied on traditional Galenic medicine while still using chemical remedies that seemed to be of practical value. See Debus, "Paracelsian Medicine."

15. On the manor, see Lipscomb, *History and Antiquities*, 4:222. Napier himself did not live or work in the manor, which was bought for his nephew, Sir Richard, in 1632.

16. For the history and architecture of the church, see Williams, *St. Andrew's Church*. The list of rectors opens with Galfridus de Gibbewin (ca. 1215) and ends with Napier.

17. Collins, *English Baronetage*, 1:225–31; Burke and Burke, *History of the Baronetcies*, 378–79. The family alias of "Sandy" (often used by Richard himself) was acquired, according to the nineteenth-century historian and family descendant Mark Napier, "from the favourite name of Alexander in the Merchiston family." Napier, *Memoirs*, 6.

18. Sir Robert Napier, 1st Baronet of Luton Hoo, Bedfordshire, was a successful merchant who conducted trade with the New World. On Sir Richard Napier, a member of the College of Physicians, see Andrews, "Napier, Richard," 183. A paper signed in 1632 records a "family settlement whereby Richard Napier alias Sandy, the Elder, of Great Linford, clerk, covenants to stand seised of all his messuages, lands, tenements, rectories, advowsons, etc., in Great Linford to the use of Richard Naper the Younger, his nephew and godson, the second son of Sir Robert Naper of Luton Hoo, Bedfordshire," Aylesbury, Centre for Buckinghamshire Studies (hereafter CBS), D-U/1/1, 1632. Both Sir Richard and his sister Mary were married into two of the region's most eminent families—Tyringham and Myddelton (or Middleton), while their eldest brother, Robert, was a frequent visitor to Great Linford. Burke and Burke, *History of the Baronetcies*, 378.

19. Ashm. 213, fol. 110.

20. Andrews, "Napier, Richard," 181.

21. Ashm. 213, fol. 110. See also Ashm. 413, fol. 63.

22. Napier was officially presented to the benefice by Edward Kempton, the husband of his sister Katherine. The advowson of the rectory was purchased in 1606 by his brother Robert. CBS, D-U/1/32, 1606–1814; Lipscomb, *History and Antiquities*, 4:225. Napier's own notes concerning his admission to the rectory are found in Ashm. 413, fol. 230v.

23. Ashm. 213, fol. 110.

24. According to his own testimony, Napier suffered throughout his life from various ailments, including "an extreme itch & scabs, . . . cholick [colic] & gout in my feete, bad legs to goe, . . . agues tert[iary] & quotidians, hed aches [headaches], gydyns deasees [giddiness?], bad sight," and was "much afflicted with mopish melancholy." Ashm. 213, fol. 110 (see also Ashm. 413, fol. 63). He first consulted Forman in September 1596 regarding some of these problems. Ashm. 234, fol. 98. The scholarship on Forman's life and work includes Kassell, *Medicine and Magic*; Rowse, *Sex and Society*; Traister, *Notorious Astrological Physician*. A long-term and fruitful relationship ensued between these two very dissimilar men. After Forman's death in 1611, Napier acquired his mentor's manuscripts from his widow, Jean, with whom he continued to be in touch. For Forman's letters to Napier, see Ashm. 240, fols. 103, 104, 106; Ashm. 1488, part 2, fol. 89.

25. Still at Lambeth, Napier attempted to practice his skills on local patients. See, for example, Ashm. 182, fols. 21v (Blage), 92v (Wilkins). Geomancy is a method of divination based on the analysis of a series of sixteen figures formed by drawing lines of dots of randomized lengths, which seems to have been used by Napier only at the outset of his medical career.

For some of Napier's geomantic calculations and notes, see Ashm. 182, fols. 178v–79, 180, 183v–84, 214, 215v–16, 219, 220v–21, 231r–v.

26. Napier counted among his clients the Earl of Sunderland, dean of Westminster (and later Lord Chancellor and archbishop of York) John Williams, the prominent pro-Calvinist theologian William Twisse, the lunatic Viscount Purbeck, and his mother, the Countess of Buckingham. See Ashm. 196, fol. 103; Ashm. 223, fols. 49v, 93; Ashm. 402, fol. 30v; Ashm. 414, fols. 15v, 192; Howell, *Epistolae Ho-Elianae*, section 5, 24 (letter XVIII). Although he was never formally educated in his new profession, Napier received a license to practice medicine, in December 1604, from Erasmus Webb, archdeacon of Buckingham. His burial was recorded in a short yet compelling entry in the parish register: "April 15, 1634. Buried, Mr. Richard Napier, rector, the most renowned physician both of body and soul." Lysons and Lysons, *Magna Britannia*, 1:597.

27. Dwelling from 1614 in the estate of Toddington, Bedfordshire, Wentworth was created Baron of Nettlestead and the first (and last) Earl of Cleveland in 1625, and later served as a royalist army officer in the Civil War. He should not be confused with his contemporary namesake, the 1st Earl of Strafford, a key figure in the personal rule of Charles I and the powerful ally of Archbishop Laud. Rutton, *Three Branches*. The seventeenth-century astrologer William Lilly testifies in his famous autobiography that thanks to his influential patron, Napier even succeeded in keeping the recusant and astrological practitioner William Marsh out of trouble. See Lilly, *Last of the Astrologers*, 48–49.

28. On Napier's manuscripts coming into Ashmole's hands, see Josten, *Elias Ashmole*, 1:210; 4:1454–55. This was sometime between Sir Richard's death on January 17, 1676, and February 1677.

29. Edward Lhuyd to Martin Lister, April 18, 1693, Oxford, in Gunther, *Early Science*, 14:177–78.

30. London, British Library, MS Sloane (hereafter Sloane).

31. Napier kept his papers unbound or in notebooks, filling the opening and closing pages of each with notes of loans made, books lent, and personal memoranda. When Ashmole bound the cases, he inserted Napier's angelic interviews from the same year at the end of each casebook. Letters and tracts were bound in clusters in separate cases. For a complete catalogue and index of these manuscripts, see Black, *Catalogue*; Macray, *Index*.

32. MacDonald, *Mystical Bedlam*. MacDonald made extensive use of the term "mental," which is now regarded as anachronistic in relation to seventeenth-century medicine.

33. See, for example, Torrey and Miller, *Invisible Plague*, 17; Parry, *Arch-Conjurer*, 264.

34. Sawyer, "Patients, Healers, and Disease."

35. This represents a major trend in the history of science that had significant implications for the histories of medicine and magic. See, for example, Shapin and Schaffer, *Leviathan*; Shapin, *Social History*. On society and magic, see Clark, *Thinking with Demons*.

36. Risse and Warner, "Reconstructing Clinical Activities," 199. Among these earlier case studies are Beier, *Sufferers and Healers*; Duden, *Woman Beneath the Skin*; Huffman, *Robert Fludd*; MacDonald, *Mystical Bedlam*.

37. Armstrong, "Patient's View"; Fissell, "Disappearance"; Porter, "Patient's View." See also the more recent Condrau, "Patient's View."

38. Pelling and Webster, "Medical Practitioners"; Pelling, *Common Lot*, "Defensive Tactics," "Knowledge Common and Acquired," and "Public and Private Dilemmas"; Wear, *Health and Healing*; Webster, *Great Instauration*, chap. 4. For a survey of unlicensed practitioners and their conflicts with the London College of Physicians, see Pelling and White, *Medical Conflicts*, esp. 136–88; and Pelling and Webster, "Medical Practitioners."

39. This might explain why only a small fraction of Napier's clients came from Great Linford itself. Offering sophisticated and often costly medical care, he seems to have been a second or higher option in the early modern "pyramid of healers."

40. On clerical practitioners in England, see Pelling and Webster, "Medical Practitioners," 199. On priestly healing, see O'Neil, "*Sacerdote ovvero strione*"; Harley, "Mental Illness."

41. Two such famous attacks were launched directly at Napier. See Cotta, *True Discovery*, 88–89. See also Hart, *KLINIKE*, 12–13; *Arraignment of Urines*, "Epistle to the Reader," A4r; Primrose, *Popular Errours*, esp. 10–18; Winthrop, *Winthrop Papers*, 1:306–7. Cotta's attack on Napier is mentioned in Lilly, *Last of the Astrologers*, 50. David Harley portrays such attacks as motivated by a pro-Calvinist distaste for confusing the separate vocations of physician and priest, rather than based on intellectual and economic protectionism. Harley, "James Hart."

42. Cook, *Decline*; Porter, *Health for Sale*. See also the more recent Jenner and Wallis, *Medicine and the Market*.

43. Gentilcore, *Healers and Healing*; Park, "Medicine and Magic"; Pelling and White, *Medical Conflicts*; Pomata, *Contracting a Cure*.

44. Andrew Wear maintains that the intense focus on "change" in the sixteenth and seventeenth centuries has distanced us from the daily routine of medical practice. Wear, *Knowledge and Practice*, esp. chaps. 2–3. Ann Geneva complains that historians have devoted very little attention to the actual practice of astrological medicine, as opposed to its intellectual content. Discussing what she has termed "the embarrassment factor," she maintains that MacDonald, "by deliberately ignoring the astrological system on which Napier based his diagnoses, prognoses, and cures, remains oblivious to the underlying categories and syntax which informed Napier's judgment and discourse." Geneva, *Astrology*, 3.

45. Grafton and Siraisi, "Between the Election"; Grafton, *Cardano's Cosmos*; Kassell, *Medicine and Magic*; Nance, *Turquet de Mayerne*; Rankin, *Panaceia's Daughters* and "Duchess, Heal Thyself"; Stein, *Negotiating the French Pox*; Trevor-Roper, *Europe's Physician*.

46. On the rise of practical astrology in the seventeenth century, see the pioneering work of Thomas, *Religion and the Decline*, esp. chap. 10; Capp, *English Almanacs*; and Curth, *English Almanacs*.

47. On medical astrology, see Akasoy, Burnett, and Yoeli-Tlalim, *Astro-Medicine*; Azzolini, *Duke and the Stars* and "Reading Health"; Curth, "Medical Content"; French, "Astrology in Medical Practice"; Harrison, "From Medical Astrology"; MacDonald, "Career of Astrological Medicine." On the decline in astrology's intellectual legitimacy, see Rutkin, "Astrology," 552–63.

48. MacDonald, "Career of Astrological Medicine," 67–68; Kassell, *Medicine and Magic*, 125–30. On the significance of clinical experience in seventeenth-century medicine, see Cook, "New Philosophy."

49. Thorndike, *History of Magic*, vols. 5 and 6, *The Sixteenth Century*; Yates, *Giordano Bruno*, "Hermetic Tradition," and *Rosicrucian Enlightenment*. For a reflective assessment of Yates's work and its reception, see Copenhaver, "Natural Magic."

50. For some of the criticism on Yates, see Copenhaver, *Magic in Western Culture*, 3–24; Farmer, *Syncretism in the West*, 115–32; Hesse, "Hermeticism and Historiography"; Merkel and Debus, *Hermeticism and the Renaissance*; Rosen, "Was Copernicus a Hermetist?"; Vickers, "Frances Yates" and "Analogy Versus Identity."

51. Webster, *Paracelsus to Newton*.

52. Thomas, *Religion and the Decline*, esp. chaps. 7, 8, 10.

53. Major works on medieval and early modern natural magic include Burnett, *Magic and Divination*; Copenhaver, "Astrology and Magic," "Natural Magic," and "Scholastic Philosophy"; Fanger, *Conjuring Spirits*; Kieckhefer, *Magic in the Middle Ages*; Klaassen, "Medieval Ritual Magic" and *Transformations of Magic*; Weill-Parot, "Astral Magic" and "Contriving Classical References."

54. A notable exception is Roos, "Magic Coins." Through an examination of various astrological sigils, drawn by the German mathematician and astronomer Julius Reichelt, Roos explains the use of numerical magical squares in producing the signs of planetary angels for these active magical objects.

55. On Dee's conversations with angels, see Clucas, "False Illuding Spirits"; Clulee, *Natural Philosophy*; Harkness, *Conversations*; Parry, *Arch-Conjurer*; Whitby, *Actions*. Some recent steps toward the study of actual angelomancy are Raymond, *Milton's Angels*; Walsham,

"Invisible Helpers"; and the collections of articles in Marshall and Walsham, *Angels in the Early Modern World*.

56. "The disenchantment of the world" is a phrase coined by Max Weber, who spoke of the eclipse of magical and animistic beliefs about nature as part of the more general process of "rationalization." Weber, "Science as a Vocation."

57. Thomas, *Religion and the Decline*, chap. 9.

58. Scribner, "Reformation."

59. Walsham, "Historiographical Reviews," 527. Cf. Martin, *On Secularization*. For a recent challenge to the division between medieval and Renaissance magic, see Klaassen, *Transformations of Magic*. For other approaches, see Allen and Rees, *Marsilio Ficino*; Perrone Compagni, "Dispersa Intentio"; Bailey, *Magic and Superstition*; Cameron, *Enchanted Europe*; Kieckhefer, "Did Magic Have a Renaissance?"

60. Napier also engaged throughout in a philosophical-theological justification of each of his practices.

61. Some examples from the vast literature on the religious atmosphere of post-Reformation England are Evans, *Problems of Authority*; Hughes, "Religious Polemic"; Hutton, "Thomas Jackson"; White, *Predestination*. See also below.

62. For the Whig school of the early twentieth century, see Merton, "Science, Technology"; Weber, *Protestant Ethic*. A major work representing the Marxist approach is Hill, *English Revolution*.

63. Elton, "High Road"; New, *Anglican and Puritan*; and the collection of articles in Russell, *Origins*.

64. The term "nonconformist" was established in England after the Act of Uniformity of 1662. However, current historical research tends to define, retrospectively, English churchmen as "nonconformist" if they violated the Act of Uniformity of 1559 or generally did not conform with the church authorities of their time.

65. Lake, "Defining Puritanism," 4. See also Lake and Questier, *Antichrist's Lewd Hat*; Lake, "Joseph Hall," "Anti-Puritanism," and "Puritanism." For the opposite approach, see Haigh, "Character of an Antipuritan"; Prior, *Defining the Jacobean Church*.

66. Napier often prayed with his patients, sometimes composing prayers for them to repeat by themselves.

67. The last three methods were used by Napier to a very limited extent and will not be discussed in this book. For a taste of his many comments on the weather, see Ashm. 182, fol. 154v; Ashm. 414, fols. 134v, 143. For certain arithmantic calculations, see Ashm. 182, fols. 22, 80v, 131v–32.

68. MacDonald, *Mystical Bedlam*, 190.

69. Sawyer, "Patients, Healers, and Disease," 193–204.

70. Wear, *Knowledge and Practice*, 3–5, 441. See also Siraisi, *Clock and the Mirror*, 68–69.

71. Maclean, "Science of Nature ," 355–56.

72. Walsham, "Historiographical Reviews," 527.

CHAPTER 1

1. Lilly, *Christian Astrology*, 288. Lilly did not always refrain from prescribing medicine, and from 1665 he began to combine his astrological practice with a medical one. Josten, *Elias Ashmole*, 1:173 and 3:1197.

2. See notes 36 and 38 in the introduction.

3. Risse and Warner, "Reconstructing Clinical Activities"; Warner, "Uses of Patient Records."

4. The various copies of Forman's medical and astrological guides are now bound and labeled as MSS Ashm. 355, 363, 389, 395, 403, and 1495 (Ashm. 403 being Napier's own copy).

For a more detailed introduction to Forman's guides on astrological medicine, see Kassell, *Medicine and Magic*, esp. 131–36.

5. A minute analysis of the full body of Napier's materials would have made for a book that was far too long. Napier documented around 4,900 medical encounters in 1602/3, preserved now in four separate casebooks: MSS Ashm. 197, 207, 221, and 404. Many of these are in the hand of Gerence James (or Games) (1568–1646), Napier's curate, who, from at least 1602, also served as an amanuensis in his medical practice. James wrote the drafts for some of Napier's letters and tracts and copied several manuscripts for his use. In 1605 he became the vicar of Pattishall, Northamptonshire, and was later admitted to the nearby parsonages of Tiffield and Paulerspury. The study of Napier's (as well as Forman's) work might be transformed by the emerging Casebooks Project, led by Lauren Kassell. This important venture promises to greatly aid future research by offering a free, digital edition of the consultations. See *Casebooks Project*.

6. Records from later periods are usually less structured, but feature the same basic components.

7. This does not mean that early modern medicine could not be practiced without reference to astrology, or that astrology was not condemned, sometimes, by members of the medical profession. See Wear, "Galen in the Renaissance."

8. Ashm. 221, fol. 280v (Barly). *Alhandal*—a purgative extract of colocynth (bitter cucumber), used as a violent purge of phlegm and choler. *Jeralog* (*Hiera Logadii*)—a purgative drug, mixing aloes and white canella, which were made into an electuary with clarified honey and kept as a dry powder; usually used to purge melancholy, but also known to help vertigo. "Decoction" is a general term referring to any preparation made by boiling herbs and then straining the water. Napier probably referred to the "common decoction for a clyster" (*decoctum commune pro clystere*), which was made from various herbs according to the humor abounding. *Pulegium* (*Mentha pulegium*)—pennyroyal, a plant in the mint genus that has been traditionally employed as a menstrual flow stimulant.

9. Cf. Pomata, "Menstruating Men"; Duden, *Woman Beneath the Skin*, 109.

10. For a comprehensive review of Galenic medicine and its history, see Temkin, *Galenism*. On Galen and Galenism in the seventeenth century, see Nutton, "Fortunes of Galen." Napier's sources on Galenic medicine remain mostly obscure. It is clear from his apologetic tract on astrology, however, that he was familiar with the astrological notions attributed to the ancient physician in the commentaries on his *De diebus decretoriis* (On Critical Days) and in the pseudo-Galenic tract *Prognostica de decubitu ex mathematica scientia* (Prognosis According to Astrology). Ashm. 242, fol. 193v.

11. Two such medical guides are Barrough, *Method of Phisick*, and Wirtzung, *General Practise*. Wirtzung discusses many diseases that were attributed to specific body organs in a head-to-foot approach, while still referring to whole-body conditions such as pain, sleep, or "cold diseases."

12. Duden, *Woman Beneath the Skin*.

13. See Wear, *Knowledge and Practice*, 134–35; and Rankin, "Duchess, Heal Thyself," 129–35.

14. Duden, *Woman Beneath the Skin*, 106–7. Shigehisa Kuriyama addresses the same problem when he discusses the paths of blood vessels on which ancient bloodletting was based, and which were not manifested in accounts of dissections. Kuriyama, "History of Bloodletting."

15. See esp. Kuriyama, "Forgotten Fear," 420–23.

16. Galen, "On the Causes of Disease" and "On the Humours." Clearly, Napier practiced seventeenth-century Galenic medicine rather than Galen's exact version of it. Still, I am using this specific edition of Galen to illustrate certain points about humoral medicine that remained basically unchanged.

17. Ashm. 221, fol. 36 (Underhill).

18. Ashm. 221, fol. 282 (Wood).

19. Ashm. 221, fol. 316v (Goodman). On the rising of body organs, see also Rankin, "Duchess, Heal Thyself," 130, 134.

20. Ashm. 403, fol. 84.

21. Ashm. 403, fol. 189v.

22. Duden, *Woman Beneath the Skin*, 106–7, 119–23. Napier seldom used the rhetoric of fluxes and inner flows, although he might have thought, like others, that a flux was responsible for the movement of disease and pain inside the body. I have been, therefore, very cautious in using such terminology here.

23. Ashm. 221, fol. 36 (Norwood).

24. Ashm. 221, fol. 133 (Byrd). Another example of a patient voiding phlegm is Ashm. 221, fol. 279 (Broughtwell). Other patients voided blood from every possible orifice: Ashm. 221, fols. 198v (Harrys), 227 (Burton and Mathew), 315v (Tyringham).

25. Pomata, *Contracting a Cure*, 129–39; Wear, *Knowledge and Practice*, 136–43; Kuriyama, "Forgotten Fear," 419–28.

26. Pomata, *Contracting a Cure*, 133. See also Pomata, "Menstruating Men." Pomata basically argues that bodily discharge, whether natural or induced, was essentially the cure for any ill state of the obstructed body.

27. Some examples of Napier's patients with stopped menses are Ashm. 221, fols. 42 (Andrewes), 179v (Walters), 224v (Earle), 283v (Briant). For patients who "cannot go to stool," see, for example, Ashm. 221, fols. 222 (Kathernes), 227 (White).

28. Kuriyama, "History of Bloodletting," 27–36; Stolberg, *Experiencing Illness*, 89–95.

29. Ashm. 221, fol. 277 (Lynnell).

30. For examples of the white flow, see, for example, Ashm. 221, fols. 277 (Rainsford), 280 (Rogers). For a case of excessive menstrual flow, see Ashm. 221, fol. 224v (Langton). Similar conditions were identified by Duden in Dr. Storch's notes, as in a case where "[a] husband has sent word that the body of his wife has opened with discharge of much . . . matter and winds." Duden, *Woman Beneath the Skin*, 83.

31. Galen, "On the Humours," 16.

32. Galen, "On the Causes of Disease," 55.

33. Kuriyama, "Forgotten Fear"; Pomata, *Contracting a Cure*, esp. 129–35.

34. Ashm. 221, fol. 222v.

35. See, for example, Ashm. 221, fols. 178v (Tarry), 278v (Tyrrell), 279 (Dorrell), 316 (Warren).

36. See Weisser, "Boils, Pushes, and Wheals."

37. Ashm. 403, fol. 190.

38. Duden, *Woman Beneath the Skin*, 109, 119–23; Weisser, "Boils, Pushes, and Wheals."

39. Pomata, *Contracting a Cure*, 1313–15.

40. On the early modern experience of disease and pain, see Stolberg, *Experiencing Illness*, 27–32, 161–212.

41. Wear, *Knowledge and Practice*, 135–36.

42. Duden, *Woman Beneath the Skin*, 111.

43. Nance, *Turquet de Mayerne*, 113–16; Stein, "Meaning of Signs," 624–39.

44. Maclean, *Logic, Signs, and Nature*, 281–301. Maclean argues that learned diagnosis in the Renaissance could not be reduced to a tree diagram, but was rather a "dialectical or endoxical division" (299).

45. Wirtzung, *General Practise*, 37.

46. Maclean, *Logic, Signs, and Nature*, 285; Wear, *Knowledge and Practice*, 134–35.

47. Wear, *Knowledge and Practice*, 134. On the perception of disease as something that moves inside the body, see also Pomata, *Contracting a Cure*, 129–30; Stein, "Meaning of Signs," 636–37.

48. Ashm. 403, fol. 192. The quote is taken from Napier's copy of Forman's astrological-medical guide.

49. Ashm. 239, fol. 2 (Lady Windsor).

50. Seitz, *Ein Nutzlich Regiment*, 1:7–29, here 12, in Stein, "Meaning of Signs," 635.

51. Cf. Duden, *Woman Beneath the Skin*, 153–57.

52. Ashm. 207, fol. 84v (Tingewicke).

53. Ashm. 221, fol. 283 (Hilpin).
54. Ashm. 221, fol. 278v (Goodman).
55. Ashm. 221, fol. 316 (Marren).
56. Ashm. 221, fol. 223v (Hollywell).
57. Ashm. 221, fol. 284 (Anglissay). See similar examples in Ashm. 207, fol. 84v (Parkyns); Ashm. 221, fols. 224 (Atkyns), 281v (Muskott), 282v (Tyrrell).
58. For such a view, mainly building on Foucault, see Armstrong, "Patient's View." However, as I will later argue, great caution should be exercised when basing any conclusions on patients' complaints in Napier's consulting room.
59. I will expand on the multiple meanings of the early modern *cause* in my discussion of the essence of *disease* below.
60. Ashm. 1495, fol. 32. The second category covered everything that did not result from inner body conditions but rather from external or environmental causes that were not divine in nature.
61. Galen, "On the Causes of Disease," 61.
62. As opposed to a sickness "of some extraordinary and outward cause, . . . not naturall by hummors or by the influence of the heavens." Ashm. 403, fol. 89v.
63. Ashm. 363, fol. 137v; Ashm. 403, fol. 89v. Forman does not explain what kind of "thoughts" can be regarded as a cause of disease, but it can be presumed that he refers to troubling or distressing thoughts.
64. Ashm. 1495, fol. 32. For a contemporary discussion of God as a cause of disease, see Burton, *Anatomy of Melancholy*, 1:159–60.
65. The early modern dichotomy between a physical and a mental disease should not be confused with our modern one. For Napier and his contemporaries, a mental disease was the situation in which the soul, basically through the passions, affected the body and disturbed its humoral balance. That a fundamental distinction nonetheless existed is confirmed by directions, such as Forman's, to know whether the disease was "in the body or in the mynde." Ashm. 403, fol. 169.
66. This is supported by much of the historical research into early modern medicine, which found that mental afflictions and externally caused harms were treated in much the same way as internal diseases. See, for example, MacDonald, *Mystical Bedlam*, 173–231; Duden, *Woman Beneath the Skin*, 140–49. Nevertheless, Napier was sometimes careful to distinguish between diseases of the body and mind. After erecting a figure for one patient, he noted, "not ill but troubled in mynd only & vnwilling to have any to know it." Ashm. 174, p. 285.
67. Duden, *Woman Beneath the Skin*, 140.
68. Quite literally, fear was therefore something bad that had to be expelled from the body. See Gentilcore, "Fear of Disease."
69. Duden, *Woman Beneath the Skin*, 140–49, here 149. Although fear as a cause of disease is very seldom mentioned by Napier, it does come up in his notes as a symptom: "was frighted in the night and divers tymes cryes out away with that blacke thing." Ashm. 221, fol. 222v (Shawe). In a rather extraordinary case, he described at length the story of a gentlewoman widow who was almost frightened to death by what was presumably her husband's spirit. Ashm. 221, fol. 223v (Crimson). See also Ashm. 221, fols. 36v (Harrys), 133v (Sparver), 134 (Richmond), 314v (Thornehill).
70. Ashm. 404, fol. 275 (Miles).
71. Webster, *Paracelsus to Newton*, 4.
72. Ashm. 242, fol. 190.
73. Ashm. 242, fol. 190v.
74. Ashm. 242, fol. 194v.
75. Chapman, "Astrological Medicine," 284.
76. Some of the most extensive and helpful vernacular guides to astrological medicine are Blagrave, *Astrological Practice*; Coley, *Clavis Astrologiae*; Culpeper, *Astrologicall Judgment* and *Semeiotica Uranica*; Eland, *Tutor to Astrologie*; Lilly, *Christian Astrology*; Salmon, *Synop-*

sis Medicinæ; Saunders, *Astrological Judgement*. A still helpful review of early modern astrological guides is Dick, "Students of Physic."

77. Capp, *English Almanacs*; Kelly, *Practical Astronomy*.

78. "Mr Barton of Eastcote," Ashm. 221, fol. 283; "I went to see Mrs Byrd," Ashm. 221, fol. 133v. Both were frequent patients.

79. "Mr Cayly sent for his daughter," Ashm. 221, fol. 280v; "Oolmans sone of Feny stratford," Ashm. 221, fol. 278; "Edward, Sir Arthur Throckmorthons servant," Ashm. 221, fol. 225; "Goodwife Easebyes childe, 3 yeeres old," Ashm. 221, fol. 222v.

80. "Goodman Ashly of Bugbrooke," Ashm. 221, fol. 225v.

81. Ashm. 1495, fol. 29. Forman devised these rules specifically for remote consultations, in which the messenger might hold back important information or even try to trick the physician.

82. It is not always clear whether a patient came to Great Linford in person or only sent his or her query by a messenger. Occasional remarks such as "was present," "sent vp her water," or "sent with a cart" suggest that both were an option. Ashm. 213, fol. 158 (Wyllyson); Ashm. 230, fol. 124v (Archpole); Ashm. 235, fols. 157v (Seabrooke), 158 (Ashwell), 161 (Write).

83. John Booker's astrological case notes lack any kind of patient identification. See Ashm. 183, pp. 1–380, and Ashm. 339, fols. 176–84. John Hall's case notes begin with the patient's title and last name only, or even with a general form of identity, such as "One Hudson" or "A certain woman of Uline." Hall, *Select Observations*. A selection of Hall's medical case notes was published in Lane, *John Hall*, and in Joseph, *Shakespeare's Son-in-Law*.

84. Ashm. 403, fol. 189.

85. More than 60 percent of Napier's patients in 1602–3 were women. Lauren Kassell found a similar proportion in Forman's practice; see Kassell, *Medicine and Magic*, 130–31. MacDonald reports an average gender ratio of 78.8 (i.e., 55.9 percent women) in four sample years (1600, 1610, 1620, and 1630); see MacDonald, *Mystical Bedlam*, 38.

86. See, for example, Ashm. 221, fols. 224 (Manners), 229 (Whitlocke), 282 (Wood), 316 (Gilbert).

87. For an example of Napier using both a decumbiture and a nativity, see the case of young Elizabeth Jennings, described in the introduction.

88. Glennie and Thrift, *Shaping the Day*, 135–80; Landes, *Revolution in Time*, esp. 85–97.

89. Ashm. 1495, fol. 29.

90. Ashm. 221, fol. 36 (Underhill).

91. For some examples of patients who came at the same time, see Ashm. 221, fols. 284v, 316v. Napier seems to have regarded the time of arrival, rather than the time of consultation, as "the time of the question asked." In one instance he explicitly states that he is casting a chart retrospectively for the time of visit, having failed to do so in the first place: "Apr[il] 17 . . . I came to sir Henry longfilds howse *quæro an evadet* 1605. [T]his motion [of erecting a chart] came not in my hed vntill this 20 of Apr[il], yet I will for experience sake see how it would fall out if I should have set it for that weddnesday which I forgote." Opposite this figure Napier composed a second chart for the time at which he documented the case, noting beside it: "this also I sett of myne owne hed, no man demanding it." Ashm. 216, fol. 55 (Longueville).

92. Some examples are Ashm. 207, fols. 36v (Ablesome), 83 (Malyn and Read), 84 (Francis), 133 (Browne).

93. See, for example, Ashm. 221, fol. 279 (Denton).

94. For example: "The fits occur a week before the moon changes and asmuch after," Ashm. 207, fol. 84 (Francis); "about the change of the moone," Ashm. 221, fol. 42 (Adcocke). On the significance of the Sun and the Moon in astrology and medicine, see Roos, "Luminaries in Medicine." As I will later show, the position of the Moon (rather than its phase) had a key role in Napier's interpretation of the astrological chart.

95. The *Lord of the day* simply signified the day of the week (☉ for Sunday, ☽ for Monday, and so on). The *Lord of the hour* was the planet signifying the *planetary hour* at the time of the visit: daylight hours were divided into twelve identical parts, called the *planetary hours* (or

artificial hours) of the day. Each part was then assigned to a planet, starting from the *Lord of the day* and advancing by order. Agrippa, *Three Books*, book 2, 371.

96. See, for example, Grafton, *Briefe Treatise*, 85.

97. The Dragon's head and tail (also known as "the Moon's nodes") are the two points at which the path of the Moon crosses the ecliptic, which is the path of the Sun through the heavens as seen from the Earth. In astrology, they are considered "shadow planets" that influence human destiny very much like regular planets, and are always in perfect opposition to each other.

98. I found *Blagrave's Astrological Practice of Physick* to be the most helpful guide for erecting a figure. Blagrave, *Astrological Practice*, esp. 28–35.

99. The Regiomontanus system was based on twelve equal divisions along the Earth's equator that were projected upon the ecliptic. This system, named after the fifteenth-century German astronomer and astrologer Johann Müller of Königsberg, whose pseudonym was Regiomontanus, underlay the most popular method for setting up the houses in early modern charts, but was largely replaced in the eighteenth century by the new Placidus system.

100. The differences between two distinct *tables of houses* are very minor even over a period of several decades. This proves that the astronomical data and calculations underlying this table did not change throughout the sixteenth and seventeenth centuries. Compare, for example, Field, *Ephemeris*, and Blagrave, *Introduction to Astrology*, 13–24.

101. The latitude of Great Linford is 52°4′, while that of London is 51°32′—a difference of only 32′, which was hardly significant for astrological calculations such as Napier's. Nevertheless, Napier was certainly aware of the difference. See his list of latitudes for different places in Ashm. 339, fol. 164v.

102. This is the only stage in the process in which we actually have a record of Napier's intermediate calculations.

103. See Ashm. 228, fols. 1r–v, for exemplary lists of *planetary hours* on three different dates in August and September 1598.

104. Siraisi, *Early Renaissance Medicine*, 124.

105. According to Claudia Stein, in the sixteenth-century French pox hospital of Augsburg a medical examination of the naked candidate was carried out by a physician and surgeon. Stein, *Negotiating the French Pox*, 136–41.

106. Ashm. 221, fols. 194v, 199v.

107. Ashm. 221, fol. 199v (Southwicke).

108. Ashm. 221, fols. 179 (Flanders), 230v (Wilson), 280 (Egerton).

109. Ellen Knight of Great Linford, visiting Napier in March 1604, seemed to him "greenish & somewhat greenish vnder the eyes as if she had the greene sickness." Ashm. 415, fol. 42 (Knight). For other examples, see Ashm. 221, fols. 281 (Lewell), 281v (Muskott), 282 (Foster).

110. Ashm. 412, fol. 224v (Tyringham).

111. Ashm. 221, fol. 284 (Rogers).

112. See, for example, Ashm. 221, fol. 316v (Goodman).

113. Brian Nance suggests that the Swiss-born physician and Napier's contemporary Turquet de Mayerne "touched the patient as a way to judge the temperature and moistness of the skin, and thus the condition of the underlying matter." Nance, *Turquet de Mayerne*, 97.

114. Ashm. 207, fol. 85 (Charles).

115. Ashm. 207, fol. 84v (Tingewicke). However, and as my discussion below suggests, it is possible that some observations about body temperature arose out of Napier's analysis of the astrological chart.

116. Ashm. 221, fol. 133 (Caddesden).

117. Stolberg, "Decline of Uroscopy."

118. Siraisi, *Early Renaissance Medicine*, 124–25.

119. Ashm. 363, fol. 5v.

120. Kassell, *Medicine and Magic*, 138.

121. Ashm. 221, fol. 282 (Ferme).

122. Sawyer, "Patients, Healers, and Disease," 247–53.
123. Ashm. 228, passim.
124. Atwell, *Apology*, 26–27.
125. In medical reports today, the patient's "main complaint" is intentionally documented independently of the doctor's "current illness" in order to clearly separate the subjective from the objective. On the illness discourse of early modern laypeople, see also Wear, *Knowledge and Practice*, 149–53.
126. Stein, *Negotiating the French Pox*, 123–25.
127. Ashm. 1495, fol. 32.
128. For example: "a pricking gnawing and shooting," Ashm. 221, fol. 133 (Caddesden); "throbbing burning shooting pain," Ashm. 221, fol. 179 (Wickens).
129. Sawyer, "Patients, Healers, and Disease," 290–93.
130. See Ashm. 355, fols. 147–51.
131. Wear, *Knowledge and Practice*, 126–30.
132. Hall, *Select Observations*, 36 (Observ. XXXIX).
133. See the introduction.
134. Sawyer, "Patients, Healers, and Disease," 193–204; Rankin, "Duchess, Heal Thyself." See also Duden, *Woman Beneath the Skin*, 72–78, 111–12.
135. See, for example, Ashm. 221, fol. 283v (Meakyns), and the case of Elizabeth Jennings, described in the introduction.
136. Ashm. 221, fol. 133 (Cutbert). Another patient wanted "to be let blood but not to take physick." Ashm. 201, fol. 142 (Stanton).
137. Ashm. 221, fol. 222v (Bradby).
138. Ashm. 221, fols. 133v (Byrd), 228 (Fairy), 278v (Tyrrell).
139. Ashm. 221, fols. 228 (Masters), 228v (Robinson), 316v (Goodman).
140. Ashm. 221, fols. 36 (Norwood), 199 (Howe), 226 (Hobson).
141. Ashm. 221, fols. 226v (Morton), 280 (Egerton), 283v (Meakyns).
142. Ashm. 363, fols. 2, 6v.
143. Forman's table of organs and astrological states appears in Ashm. 355, fol. 153. For other versions of this table, see Eland, *Tutor to Astrologie*, 40–41; Lilly, *Christian Astrology*, 119–20; Saunders, *Astrological Judgement*, part 1, 37–38.
144. See, for example, Ashm. 403, fols. 189–95. One such rule advised, "♂ 29 [degrees into] ♉ d[ominus] 12 [and the Moon separating from] ☉ 9 [degrees into] ♌, ☿ [in] ♌ *[dominus hora]*, causeth much payne of the head and stomacke, . . . much payne of the belly, in danger of yellow jaundise, hott in the upper parts of the body and colde in the feet, and some fluxe, either of the whites or laske." Ashm. 403, fol. 189v.
145. Ashm. 242, fol. 194v. Napier refers to "a peculiar treatise of Astrology" by "Arnoldus denovavilla," which is probably the pseudo-Arnaldus astrological treatise *De Sigillis*, wrongly attributed to the Catalan physician Arnaldus de Villa Nova (1240–1311).
146. Kassell, *Medicine and Magic*, 141–43. Kassell supports her hypothesis with several examples of adjacent cases that list similar symptoms.
147. Ashm. 207, fol. 84 (Harte)—eight different complaints; Ashm. 221, fol. 194v (Coxes)—pain at eight different body parts; Ashm. 221, fol. 315 (Cooke)—eight different complaints.
148. Ashm. 221, fol. 280 (Fitzhughe).
149. Ashm. 221, fols. 280v (Barly), 281 (Lewell), 283 (Deb); Ashm. 414, fol. 131v (Sealy). Still, drawing meaning from seventeenth-century language might be risky, as the verb "to lie" might have referred to pain. See, for example: "A payne in his left side lyes heavy and so rises to his heart and chest." Ashm. 221, fol. 315v (Young).
150. Ashm. 221, fol. 224 (Brinklowe).
151. Ashm. 221, fol. 36 (Underhill). "Stomach ventosity" refers to a flatulent (or "windy") stomach.
152. How the *astrological cause* and signifiers of disease were determined, I will explain below in my discussion of how Napier identified the underlying problem.

153. Ashm. 221, fol. 280 (Fitzhughe).

154. Ashm. 221, fols. 226, 228v (Robinson).

155. In the first case, the case notes are not accompanied by a separate chart, and Napier probably used one that was drawn for another patient who consulted him half an hour before.

156. Ashm. 221, fol. 282 (Ferme).

157. Ashm. 221, fol. 279 (Myddleton). In the case of "a godly religious honest man" from Hanslope, who was "afflicted in mynd & tempted," Napier noted that "he would not tell his infirmitye." Ashm. 415, fol. 143 (Harburke).

158. This whole discussion emphasizes the already acknowledged danger in relying only on doctors' notes when reconstructing past diseases or even patients' complaints and feelings. It can also give a new twist to the historiographic discussion of the change of focus from the patient narrative to medical expertise and observational givens.

159. The planet that is Lord of the sign that was on the cusp of the sixth house.

160. Saunders, *Astrological Judgement*, part 2, 8.

161. Forman expands upon this step in the process of astrological judgment in Ashm. 403, fols. 81, 84v.

162. Some rather late examples (from 1602–3) in which Napier indicates the astrological cause are Ashm. 207, fol. 85 (Tinsly); Ashm. 221, fols. 283v (Trussell), 315 (Herryng).

163. Ashm. 403, fol. 81. It is specifically of this method that Forman boasts, "this is a great secret, for never any man before my tyme that I could read or heare of, did fynde this secret nor the secrets of the Judgments of diseases, nor shewe howe to know the causes in his degrees of every disease and sicke person. But God that is the only giver of wisedome and knowledge hath given me the true knowledge thereof" (ibid.).

164. The best example of a guide listing such rules is probably Richard Saunders's *The Astrological Judgement*, esp. part 1, 114–208. Saunders was a member of William Lilly's circle and acted as physician to Lilly and Elias Ashmole. Certain sections of his book are an exact or approximate copy of Forman's instructions, to which he was probably exposed through his acquaintance with Ashmole's library. This makes him an even better source for my purpose. Forman himself refers to a similar set of instructions in "my booke of the 7 pl[anets] [in which it] is shewed what disease every pl[anet] doth cause throughout the xij [12] signes at large, if he be L[ord] of the 6 or 12 house." Ashm. 403, fol. 81.

165. Ashm. 403, fol. 84v.

166. My analysis is based on Saunders, *Astrological Judgement*, part 1, 16–19, 67–68, 86, 158–59.

167. See, for example, Ashm. 221, fols. 36v (Norwood), 179v and 199 (Sanders), 283v (Trussell). In later periods, such detailed and complex analysis of the chart is much more frequent and more extensive. See, for example, Ashm. 237, fol. 37 (Hadderick); Ashm. 413, fol. 105v (Longsvill).

168. Ashm. 403, fols. 189–93v. Most of these rules relate to a combination of the Lord of the sixth (or twelfth) house, the planet from which the Moon last separated, and the Lords of the day and hour.

169. Saunders, *Astrological Judgement*, part 1, 98, 208; and part 2, 77–79.

170. On Napier correcting the astrological significators and noting later understandings of the chart, see Sawyer, "Patients, Healers, and Disease," 297–99; and below in this chapter.

171. Many such rules appear in Ashm. 403, fols. 83, 85–88v, and Ashm. 363, fols. 137v–38.

172. Ashm. 403, fol. 89v.

173. During the first part of his career, Napier almost entirely avoided making any kind of prognosis. This partly changed in the early 1610s, when he started to consult the archangel Raphael in some of his medical cases. See chapter 3 on Napier's conversations with angels.

174. Saunders, *Astrological Judgement*, part 1, 112.

175. Ibid.

176. San Marino, Calif., Huntington Library, MS Stowe (hereafter Stowe) 1500. Napier's letter is addressed to Sir William Andrews (or Andrewes) of Lathbury (b. 1590), High Sher-

iff of Bucks in 1630. Sir William's wife, Anne, was the daughter of Sir Thomas Temple (1567–1637), an English landowner and member of Parliament. Both families were regular visitors to Napier's consulting room and appear frequently in his queries to the archangel Raphael. See, for example, Ashm. 213, fols. 185r–v, 187, 189v; Ashm. 235, fols. 187–88, 191v, 192v; Ashm. 414, fol. 220v. A personal prescription, prepared for Sir Andrews, is in Ashm. 1488, part 2, fol. 29.

177. Napier referred to patients' complexions only rarely. See, for example, Ashm. 221, fol. 278v (Fisher). On the shift from records of cures to treating individuals, see Crisciani, "Histories"; Pomata, *Praxis Historialis."*

178. Wilson, "History of Disease," 275, 303–6; Pickstone, *Ways of Knowing*, 33–59.

179. Cunningham, "Identifying Diseases." Margaret Pelling makes the case for reading the history of medicine in terms of modern issues in the introduction to her collection of articles on social, demographic, and occupational diversity in seventeenth-century medicine. Pelling, *Common Lot*, 1–16.

180. See "Classification of Diseases."

181. A named disease could be "cataract" or "dropsy"; an unnamed disease could be "the maladies in the eye proceeding of choler" or "inflammation of the stomach." See Barrough, *Method of Phisick*, 51, 57, 107; Wirtzung, *General Practise*, 401.

182. King, "What Is Disease?"

183. See Sawyer's comprehensive discussion of diseases in Napier's medical practice in "Patients, Healers, and Disease," 344–524.

184. See esp. Siraisi, *Clock and the Mirror*, 195–213; Wear, *Knowledge and Practice*, 143–46.

185. Stein, *Negotiating the French Pox*, esp. 175–78, here 176. See also Siraisi, "Disease and Symptom."

186. For the use of "cause" in early modern learned medicine, see, for example, Barrough, *Method of Phisick*. Barrough opens his discussion of each disease with a list of its external triggers ("the cause"). Turquet de Mayerne concludes his judgment for a patient with a history of kidney stones and difficulty in voiding urine with "This malady began from bouncy horseback-riding." Nance, *Turquet de Mayerne*, 165.

187. Siraisi, *Clock and the Mirror*, 195–213.

188. Ashm. 221, fols. 133 (Caddesden and Browne), 225 (Johnson), 230 (Darling), 281v (Edwards), 283 (Deb).

189. Ashm. 207, fol. 84v (Parkyns); Ashm. 221, fols. 36v (Harrys), 133v (Larvyes and Sparver), 281 (Lewell), 314v (Thornehill).

190. Ashm. 221, fols. 42 (Bly), 223v (Bradbourne), 278 (Chibnoll), 315 (Cooke).

191. Wear, *Knowledge and Practice*, 143–44.

192. Forman and Napier usually noted a specific astrological cause, such as "it is caused of ♃ in ♏." More generally, these judgments involved the specification of several malevolent or fortunate planets, their aspects to other planets, and their places in the zodiac.

193. Ashm. 403, fol. 88v.

194. Forman implies the same when he concludes in one of his demonstrative examples, "not sicke, but hath a flux of nature." Ashm. 363, fol. 171v. See also Rankin, "Duchess, Heal Thyself," 129.

195. Still, Napier sometimes prepared prescriptions for named diseases. See, for example, Ashm. 1386, fol. 231; Ashm. 1457, fols. 18, 68; Ashm. 1488, part 2, fol. 10v.

196. Nance, *Turquet de Mayerne*, 105–21.

197. "Formal language," a term mainly used in mathematics, computer science, and linguistics, is defined as a set of strings of symbols that are constrained by specific rules. In certain applications, a given formal language is based on a "vocabulary": a finite set of symbols (or "words"), from which "well-formed formulas" (or "sentences") may be formed. In the case of Napier's diagnostic process, a vocabulary of astrological symbols and a few fixed strings was used to construct a formal language consisting of two well-formed formulas: "It is of ω in ω" and "The Moon separating from ω in ω."

198. Still, as I have already pointed out, contemporary medical treatises often categorized afflictions by name rather than by cause.

199. Ashm. 363, fol. 132.

200. For some personal prescriptions, see Ashm. 204, fols. 112–14; Ashm. 243, fol. 47; Ashm. 1488, part 2, fol. 29. These prescriptions differ significantly from those incorporated in the case notes, and usually include a much longer list of drugs and very detailed instructions for preparation and administration. On Forman's documentation of his treatment plans, see Kassell, *Medicine and Magic*, 152–59.

201. Pomata, *Contracting a Cure*, 129–39; Wear, *Knowledge and Practice*, 136–41; Kuriyama, "Forgotten Fear." For the use of purgation in contemporary medical handbooks and case records, see, for example, Wirtzung, *General Practise*; Pechey, *Plain Introduction*; Hall, *Select Observations*. For instructions on purgation in handbooks of astrological medicine, see Blagrave, *Astrological Practice*, 91–93; Salmon, *Synopsis Medicinæ*, 524–27.

202. Siraisi, *Early Renaissance Medicine*, 145.

203. Saunders, *Astrological Judgement*.

204. Pomata, *Contracting a Cure*, 132.

205. Salmon, *Synopsis Medicinæ*, 372.

206. Culpeper, "Key to Galen's Method," 394.

207. It was contested whether they accomplished this by their manifest qualities (hot, cold, dry, or moist) or by some hidden virtues. For a discussion of the mechanics of purgation, see Bacon, *Works*, 1:146–48.

208. The direction in which the humor was evacuated was usually determined by the kind of medicine given. The famous Renaissance scholar Marsilio Ficino noted that when using the herb hellebore the direction of the evacuation is determined by the direction in which the leaves were pulled upon when collecting the herb. Ficino, *Three Books*, 325.

209. A detailed categorization of medicines, based on these functions, can be found in Salmon, *Synopsis Medicinæ*, 360–71.

210. Ibid., 372; see also 525; Burton, *Anatomy of Melancholy*, 2:190.

211. Saunders, *Astrological Judgement*, part 2, 90. For a contemporary comprehensive discussion of preparatives (also called "concoctions"), see Hart, *KLINIKE*, 276–86.

212. Saunders, *Astrological Judgement*, part 1, 116; see also part 2, 90.

213. Burton, *Anatomy of Melancholy*, 2:190.

214. See, for example, Ashm. 411, fol. 165 (Monford).

215. Ashm. 221, fol. 279 (Dorrell). A *lohoch* is a medicine designed to be licked up with the tongue that was made by boiling the medicinal herbs in sugar, reducing them to a thick syrup. Culpeper, "Directions for Making Syrups," 208.

216. Ashm. 221, fol. 281v (Pomred).

217. Saunders, *Astrological Judgement*, part 1, 174, 181.

218. Culpeper, "Catalogue of Simples," 288.

219. Saunders, *Astrological Judgement*, part 2, 131.

220. Ibid., part 2, 91.

221. Ashm. 221, fol. 36 (Underhill). The abundance of phlegm probably refers here to a symptomatic mucus manifested by the patient, rather than to the humor that goes by the same name.

222. Ashm. 355, p. 91; Ashm. 221, fols. 278v and 283 (Barton).

223. Ashm. 221, fol. 284 (Benson and Thorneborough). The quote on hyssop is taken from Culpeper, *English Physitian*, 151. On the soothing qualities of *diaphænicon*, see Burton, *Anatomy of Melancholy*, 2:190. Further examples of its use in Napier's practice are in Ashm. 207, fol. 84v (Okeley); Ashm. 221, fol. 36v (Ablesome).

224. A medicine mixed with sugar and water or honey into a sweet pasty concoction.

225. Some cases in which Napier prescribes a syrup of erratic poppies are Ashm. 221, fols. 178v (Tarry), 179v (Philip), 225v (Langton), 229 (Britten), 230 (Farren). The main ingredient of this syrup was poppy, of which Culpeper wrote, "the garden poppy heads, with seeds made

into a syrup, is frequently, and to good effect used to procure rest and sleep in the sick and weak." Culpeper, *English Physitian*, 196. For Napier's use of lettuce (*lactuca sativa*), see Ashm. 221, fols. 179v (Philip), 198v (Hooton), 280 (Fountyne?). According to Culpeper, "the juyce of *lettuce* mixed or boyled with oyl of roses, and applied to the fore head and temples, procureth sleep." Culpeper, *English Physitian*, 141–42. For Napier's use of *diascordium*, see Ashm. 221, fols. 179v (Philip), 225v (Langton), 280 (Fountyne?). See also Culpeper, "Catalogue of Simples," 328–29.

226. Ashm. 221, fols. 178v (Stratton), 221 (Brough).
227. Saunders, *Astrological Judgement*, part 2, 48.
228. See the two letters from Thomas White (presumably the apothecary of the University of Oxford) to Napier, dated June 14, 1600, and August 6, 1602, with lists of drugs bought and their prices (Ashm. 1488, part 2, fol. 65; Ashm. 177, fol. 128v), and Nicholas Leate's account of drugs sold to Napier and their prices, from November 1601 to April 1604 (Ashm. 181, fols. 87v–88). For the appropriate times to collect medicinal herbs, see, for example, Blagrave, *Astrological Practice*, 10–12.
229. Wear, *Knowledge and Practice*, 140–41; Duden, *Woman Beneath the Skin*, 123. For nonastrological instructions on manipulating the humors to a specific place in the body, see, for example, Wirtzung, *General Practise*, 43.
230. Capp, *English Almanacs*, 205.
231. Ashm. 242, fol. 194.
232. Saunders, *Astrological Judgement*, part 2, 54.
233. Ibid.
234. Ashm. 403, fol. 177.
235. See, for example, Ashm. 221, fols. 223 (Kolte), 279 (Dorrell); Ashm. 235, fol. 160 (Whitlerke).
236. Wear, *Knowledge and Practice*, 67–82. See also Nance, *Turquet de Mayerne*, 155, 160–61.
237. Burton, *Anatomy of Melancholy*, 2:190.
238. For Forman's lists, see Ashm. 355, pp. 15–147.
239. For a list of simples appropriate to the various body parts, see Burton, *Anatomy of Melancholy*, 2:168; Culpeper, "Catalogue of Simples," 262–63. For an ascription of medicinal herbs to their governing planets, see Blagrave, *Astrological Practice*, 1–4.
240. See Ashm. 221, fol. 222 (Piggott), with eighteen different remedies prescribed in a single consultation.
241. Debus, *English Paracelsians*, 145–56.
242. Burton, *Anatomy of Melancholy*, 2:179. For Napier's use of antimony, see, for example, Ashm. 221, fols. 42 (Barkwell), 51v (Barnes), 133 (Cutbert). Antimony remained one of Napier's favorites throughout his entire career.
243. On the development and influence of Paracelsian medicine, see Debus, *English Paracelsians* and *French Paracelsians*; Grell, *Paracelsus*.
244. Ashm. 221, fol. 224 (Brinklowe).
245. Ashm. 221, fol. 281 (Key).
246. Ashm. 221, fol. 282 (Foster). Other examples of Napier returning to his notes to record the outcome of a case are Ashm. 221, fols. 6 (Coles), 225v (Langton), 230 (Farren), 277v (Bloud), 279 (Myddleton), 281 (Conyngham and Lewell).
247. For a discussion of Forman's follow-up patterns, see Kassell, *Medicine and Magic*, 137–41. For Napier's references to previous visits, see, for example, Ashm. 221, fols. 51v, 227v, and 282 (Parret) and fols. 221v, 222, and 227v (Piggot).
248. Although sometimes this was the nature of his follow-up. See, for example, a list of fits made by a patient's mother in Ashm. 1488, part 2, fol. 30.
249. Ashm. 221, fols. 225 and 228 (Johnson).
250. Ashm. 221, fol. 278 (Willymot). For other examples, see Ashm. 221, fols. 133v (Byrd), 211 (Ball).

251. Hall, *Select Observations*, 72 (Observ. LXXXI); see also 8 (Observ. VI), 87 (Observs. II and III), 95 (Observ. X). For Forman's remarks on his patients' stools, see, for example, Ashm. 355, fol. 39.

252. The only significant exception being questions of pregnancy, which are in a way medical questions. For comparison, 8 percent of Forman's querents posed nonmedical questions. See Kassell, *Medicine and Magic*, 130.

253. At the very beginning of his career (that is, until about 1600), Napier did sometimes calculate the critical days and crisis of the illness.

254. Napier's holograph copy is bound in Ashm. 204, fols. 50–63, and Ashm. 240, fols. 137–38. A more legible yet partial copy is in Ashm. 242, fols. 187–96.

255. Ashm. 389, pp. 1–879. On this treatise, see Kassell, *Medicine and Magic*, 61–72.

256. Ashm. 242, fol. 192.

257. Ashm. 242, fols. 194v–95. See Ficino, *Three Books*, 241.

258. See my discussion in chapter 4 on religion and knowledge.

259. Thomas, *Religion and the Decline*, 330–32.

260. Sawyer, *Patients, Healers, and Disease*, 294–303.

261. MacDonald, *Mystical Bedlam*, 183, 194–95; Thomas, *Religion and the Decline*, 327–30.

262. Bylebyl, "Renaissance Clinic," 51.

263. Nance, *Turquet de Mayerne*, 98.

264. Thomas, *Religion and the Decline*, 287.

265. On medicine as a conjectural "low-science," see Hacking, *Emergence of Probability*, 39–48, and the preface to the 2nd edition (2006). About the history of scientific objectivity, see Daston and Galison, *Objectivity*.

266. On the place and importance of "matters of fact" in the seventeenth century, especially in a medical context, see Cook, *Matters of Exchange*, "Victories for Empiricism," and "Physicians and Natural History." See also Shapiro, *Culture of Fact*.

267. Wear, *Knowledge and Practice*, 133–36. On observation and scientific proof, see also the collection of articles in Daston and Lunbeck, *Histories*, esp. part 2.

268. Ashm. 403, fols. 82r–v.

269. See Forman's letter to Napier, from September 1599, requesting him to hasten his promised answer "to all invectives against our profession." Ashm. 240, fol. 103.

270. See note 254 above.

271. Ashm. 242, fol. 187.

272. Napier invokes the authority of a painstakingly long list of ancients and moderns, including the holy patriarchs; great ancients such as Hippocrates, Plato, Aristotle, Ptolemy, and Galen; eminent rulers such as Alexander the Great, Julius Caesar, Charlemagne, and Alphonso X of Castile; Christian fathers such as Basil of Caesarea, Ambrose, Augustine, and Pseudo-Dionysius the Areopagite; Muslim and Jewish scholars such as Avicenna, Avenzoar, and Maimonides; medieval Christian luminaries such as Albertus Magnus, Bonaventure, Thomas Aquinas, and Duns Scotus; Reformed theologians such as Martin Luther and Lambert Daneau; and famed Renaissance figures such as Marsilio Ficino and Girolamo Cardano.

273. Ashm. 242, fol. 192.

274. Ashm. 242, fol. 190v.

275. Ashm. 242, fol. 192.

276. Ashm. 242, fol. 191v.

277. Ashm. 242, fol. 191.

278. Ibid. Napier does not hesitate to quote Hyperius on another subject where his views do match his own. See Ashm. 1148, pp. 206, 209, 220. A much broader discussion of Napier's views in this matter appears in chapter 4.

279. Ashm. 242, fol. 190.

280. Ashm. 242, fol. 192.

CHAPTER 2

1. MacDonald, *Mystical Bedlam*, 13–32, here 19. See also Sawyer, "Patients, Healers, and Disease," 315–37; "Strangely Handled." One of the earliest sources depicting Napier as a master of magical arts is John Aubrey's *Brief Lives* (originally published as *Minutes of Lives*), written between 1669 and 1693.

2. Lilly, *Last of the Astrologers*, 49. Frank Klaassen recites a somewhat similar story, involving a thirteenth-century apothecary who received a magical image of a ship from the celebrated astrologer Guido Bonattim and destroyed it on the advice of a priest. Examining the meaning of such morality tales, Klaassen contextualizes them under the heading "The Apothecary's Dilemma." Klaassen, *Transformations of Magic*, part 1.

3. Ashm. 421, fol. 158.

4. Black, *Catalogue*, 327. MS Ashm. 182 contains, according to Black, Napier's "prayers and exorcisms for the sudden deliverance of persons possessed with evil spirits," but these are not necessarily texts of ritual dispossession and therefore provide no further evidence. See Ashm. 182, fols. 167v–70.

5. "Ague" is an early modern disease-name, describing a state of intermittent fever. For Napier's terminology of fevers, see Sawyer, "Patients, Healers, and Disease," 360–61.

6. Aubrey, *Brief Lives*, 217. Both this and the chapter's opening quotation reveal the practical face of seventeenth-century English religious conflicts, with which I will engage in chapter 4. At the same time, they also testify to Napier's emerging reputation as a magus.

7. One remarkable exception is a short text titled "Against witchcraft," written in Napier's hand, outlining a ceremony that involved putting needles in the patient's heated urine while citing the Trinitarian formula. Add. 36674, fol. 145. However, there is no evidence to support the claim that Napier ever used such a ritual in practice. See also a charm against the plague, "Sold by the prestes [priests] at Vienna," communicated to Napier by Sir Francis Beaumont, Ashm. 1473, fol. 680; a prayer or blessing against "all harm," Ashm. 1790, fol. 113v; a charm "To heale any wound or sore, to asswage any swelling, and to cure any dangerous sore whatsoever," Ashm. 1790, fol. 120; and another "For bewitching, forespeaking or falling sickness, or to breake any Impostume or bile," Ashm. 1790, fol. 121. The last two seem to be in Gerence James's hand. In March 1622 Napier consulted the archangel Raphael about a "prayer agaynst w[itchery] & sorc[ery] in the end of my booke." Ashm. 223, fol. 194v.

8. Although Napier's angelic helpers sometimes advised him on particular remedies, the outcome of his conferences was always *information* rather than *action* (i.e., intervening in the world to actually help cure someone).

9. In this I follow the distinction laid out by the Jesuit scholar Martín Del Río at the turn of the seventeenth century with regard to artificial magic. Del Río talked of three efficient causes of magic—natural, artificial, and diabolic—and then divided the second into operative and divinatory. Operative magic is divided again into "mathematical" and "deceitful," and Del Río places in the domain of the latter the kinds of magic I discuss in this chapter, including astrological talismans. Del Río, *Disquisitionum Magicarum*, 32.

10. Lauren Kassell has reached a very similar conclusion regarding Simon Forman, Napier's tutor, maintaining that "magical activities were indivisible from Forman's work as an astrologer-physician." See Kassell, *Medicine and Magic*, 211–13, here 213.

11. Park, "Medicine and Magic," 129–49, here 138.

12. O'Neil, "*Sacerdote ovvero strione*," 53–83.

13. Copenhaver, "Scholastic Philosophy"; Weill-Parot, "Astral Magic."

14. Klaassen, "English Manuscripts," "Medieval Ritual Magic," and *Transformations of Magic*; Page, "Image-Magic." This modern categorization is primarily functional, whereas the contemporary one was ontological, differentiating between sentient and nonsentient targets of the magical act. Nevertheless, many other kinds of categorizations exist in the study of magic.

15. Klaassen, "Medieval Ritual Magic," 172.
16. Page, "Image-Magic," 75–78.
17. Pingree, "Learned Magic," 42–43.
18. Saif, *Arabic Influences*.
19. Pingree, "Diffusion"; "Learned Magic," 42, 50–54; Weill-Parot, "Astral Magic," 167–74. An early, yet still helpful, review of the literature of astrological talismans is Thorndike, "Medieval Tracts." For a more recent discussion on the origins and development of Christian astral magic, see Weill-Parot, "Contriving."
20. Saif, *Arabic Influences*, 95–123.
21. Hermetic writings were already known in medieval Europe, based on the *Corpus Hermeticum*: a group of religious and philosophical Greek texts from the early Christian era that were once believed to encapsulate the thought of an Egyptian sage by the name of Hermes Trismegistus. Their intellectual influence dramatically increased when they were translated into Latin and published in 1471 by Ficino. Yates, *Giordano Bruno*, 44–61; Ebeling, *Secret History*; Merkel and Debus, *Hermeticism and the Renaissance*. For the problematic state of "hermeticism," see note 50 in the introduction.
22. Some relatively recent works from the vast literature on Christian Kabbala are Dan, *Christian Kabbalah* and *Kabbalah*; Beitchman, *Alchemy*.
23. Ashm. 1491 and 1494.
24. Sloane 3822, fols. 68–76v.
25. Ashm. 354, fols. 5–170.
26. A more comprehensive survey of Forman's contributions to the various magical traditions can be found in Kassell, *Medicine and Magic*, 210–15, 224; "Economy of Magic."
27. These lists are scattered throughout Napier's practice books. See, for example, Ashm. 197, fol. 1; Ashm. 199, fol. 193; Ashm. 220, fol. 189v; Ashm. 237, fol. 193v.
28. Davies, *Grimoires*, 63. For Napier's extracts from *The Picatrix*, see Sloane 3679. Some of Forman's explanations about sigils are also headed "Out of *Picatrix*." Ashm. 431, fol. 146v.
29. Ashm. 407, fol. 150. On Forman acquiring this book, see Kassell, *Medicine and Magic*, 215.
30. Ashm. 213, fol. 179.
31. Napier refers to this book in a letter. Ashm. 421, fol. 166. For a manuscript copy, titled *Liber Balamini sapientis de sigillis planetarum*, see Sloane 3848, fols. 52–58. A translated copy, partly written in Ashmole's hand, is in Ashm. 188, art. V. See also Thorndike, *History of Magic*, 2:243.
32. Ashm. 242, fols. 193v, 195.
33. See, for example, Sloane 3822, fols. 22, 23, 25, 44v.
34. Sir Thomas Myddleton (also Myddelton or Middleton), the son of Sir Thomas Myddleton, Lord Mayor of London in 1613, was married to Napier's sister Mary.
35. Ashm. 431, fol. 152.
36. Ashmole even made the effort of compiling Napier's guidelines for the usage of various sigils. See note 68 below.
37. Forman strove to make his divinely inspired wisdom publicly known, yet, frustrated by mocking reactions, he usually abstained from publicizing the "secrets" of his art.
38. Ashm. 240, fol. 106.
39. Ibid. Cf. Ashm. 363, fols. 69v–71v.
40. Ashm. 392, fol. 46.
41. On Forman's use of astral magic, see Kassell, *Medicine and Magic*, 221–25; and Kassell, "Economy of Magic."
42. Roos, "Luminaries in Medicine," 447–56; Kieckhefer, *Magic in the Middle Ages*, 144–47.
43. Galen, *De theriaca ad Pisonem*, quoted in Copenhaver, "Scholastic Philosophy," 525. On the use of occult forces in ancient and medieval medicine, see Page, *Magic in Medieval Manuscripts*, chap. 2; Copenhaver, "Tale of Two Fishes." The notion of occult powers appears

implicitly in the New Testament, when simple cloths that were charged with the power of Paul the Apostle become effective talismans. Acts 19:11-12.

44. Copenhaver, "Scholastic Philosophy," 540.

45. Aquinas, *Occultis Operibus Naturae*. On Thomas's views, see Kieckhefer, *Magic in the Middle Ages*, 130–31; "Specific Rationality." This was part of a much larger debate over the various aspects of astrology. See, for example, Niccoli, *Prophecy and People*; Smoller, *History*. Astrological talismans were already widely used in the Jewish and Christian communities of southwestern Europe as early as the thirteenth century. See Zimmels, *Magicians*, 137–39; Shatzmiller, "In Search."

46. For the picture of the skies in medieval cosmology, see Grant, *Planets*. On occult qualities in early modern Europe, see Hutchison, "What Happened?"; MacDonald Ross, "Occultism and Philosophy." For Robert Fludd, the Rosicrucian physician and astrologer who was Napier's contemporary, occult metaphysics and the mechanisms of divinatory arts were "simply the way God's spirit emanated through and worked in the upper and lower realms." Huffman, *Robert Fludd*, 3.

47. Ficino, *Three Books*, 299, 301. Napier was well acquainted with Ficino's work, quoting from the third part of his *De Vita*, titled "*Coelitus Comparanda*" ("On harmonising your life with the heavens"), in his "Defence of Astrology." Ashm. 242, fols. 194v–95. For Forman's notes taken out of the same Ficinian treatise, see Ashm. 244, fols. 106–8; Ashm. 206, fol. 22. Since my primary aim is to explore the theoretical underpinnings of Napier's practice, and even though his sources of inspiration are not entirely clear, it seemed to me fitting to quote Ficino extensively on the mechanics and theoretical considerations of astral magic.

48. French, *John Dee*, 92–95. See also Clulee, *Natural Philosophy*, esp. 39–73; Harkness, *Conversations*, 71–77.

49. Ficino, *Three Books*, 305.

50. Ibid., 327.

51. Ibid., 309; see also 327.

52. Copenhaver, "Scholastic Philosophy," 538.

53. Ficino, *Three Books*, 329, 331, here 331.

54. Ibid., 327. Ficino argued that the celestial influence is acquired "through the heating produced by hammering" (343).

55. "Substantial form" and "whole substance" are originally Aristotelian concepts; I refer here to their Thomist interpretation.

56. Copenhaver, "Astrology and Magic," 274–85; "Scholastic Philosophy," 539–54; and "Natural Magic," 270–75.

57. Angus Fletcher writes of an "allegorical causation" that governed the reciprocal influence between two entities that shared such a likeness. Fletcher, *Allegory*, 181–219. Based on Ficino, Copenhaver argues that "the magus can dispose earthly objects to receive celestial powers by manipulating material species, i.e., species in a taxonomical sense, because such species are part of a hierarchy of forms that reaches through the heavens to the divine mind." Copenhaver, "Renaissance Magic," 353.

58. Ficino, *Three Books*, 333. See also Pingree, "Learned Magic," 42–43.

59. Hiebner, *Mysterium Sigillorum*, 162. I rely here on Hiebner as one of the later, more coherent writers on sigils.

60. Kassell, "Economy of Magic," 47.

61. MS Oxford, Corpus Christi 125, fol. 109, quoted in Page, "Image-Magic," 76. This is a fragment of a short work in Latin titled *Glossulae super Librum imaginum lunae*, written by an unknown fifteenth-century hand.

62. Goldwater, *Mercury*, 25–26.

63. Hiebner, *Mysterium Sigillorum*, 161.

64. Sloane 3826, fol. 93v.

65. Sloane 3822, fol. 19.

66. See note 28 in the introduction. Ashmole also received from Thomas Napier "a silver ring against the falling sicknes" that belonged to his great uncle, and "a ring made by Dr Forman in Silver." Sloane 3822, fol. 162; Sloane 3846, fol. 102v. He later made use of them in treating his own patients. See Josten, *Elias Ashmole*, 1:235 and 4:1640.

67. One of these patients was Ashmole's own wife. Josten, *Elias Ashmole*, 1:212–13; see also 226–27. Using astrological sigils was not new to Ashmole, who had already cast such objects (some for medical purposes) at least as early as 1650. Ibid., 2:537 and n3, 545–49, 584. His directions for the making of certain sigils was published around the same time in Ashmole, *Theatrum Chemicum*, 463–65.

68. Ashm. 421, fols. 171r–v. See also Ashmole's instructions on how to make sigils "[t]o helpe one that hath the Falling Sickness," Sloane 3822, fol. 20v. A very similar set of instructions on the use of astral talismans of various sorts, built on the diaries of John Dee and his son Arthur for the years 1591–92, is in Ashm. 1790, fols. 34–35.

69. As can be seen in figure 4, Napier's prescriptions of astrological talismans are easily identifiable, yet they lack any illustrations or explanatory remarks.

70. Some laminas were rather square, or were made of wood or leather; they usually served a medical purpose. Ficino commented that contemporary opinion required sigils to have a round shape, resembling the heavens, but admitted that more ancient authorities preferred to use cross-shaped sigils, which more than any other shape "possessed length and breadth." Ficino, *Three Books*, 335.

71. Sloane 3846, fol. 45v.

72. Ibid. Ficino described a sigil made of marble, impressed on some medicinal substance that was then given to the patient to swallow. Ficino, *Three Books*, 337. See also Napier's report of "Mr Wallys," who "made with olibanum pounded & made to a cake with water of St Johns wort & a litle syrupe of roses, some 6, & stamped them." Ashm. 414, fol. 138v.

73. Ashm. 421, fol. 171. Ashmole does not mention a source for this information.

74. Each planet traditionally possessed both a spirit and a planetary intelligence (sometimes called a "daemon"), respectively responsible for its baleful and beneficial influences.

75. Calder, "Magic Squares"; Lidaka, "Book of Angels," 35–36; Roos, "Magic Coins."

76. Ashm. 421, fol. 171v. Napier supposedly received this information from the archangel Raphael. His original question and the angel's response are in Ashm. 414, fol. 220.

77. Ashmole once testified, nevertheless, that Sir Richard Napier told him about a silver ring that his uncle had cast at an astrologically propitious time, on which the words "+ Dabi + Habi + Haber + Hebr +"—a popular magical formula for curing the falling sickness—were engraved. Sloane 3846, fol. 102. Napier's original directions for making such sigils seem to appear on the bottom of Add. 36674, fol. 145. Cf. Evans, *Magical Jewels*, 124; Wecker, *Eighteen Books*, 51; Harms, Clark, and Peterson, *Book of Oberon*, 146.

78. Nevertheless, Napier seems to have cast some of the sigils himself. In a note from September 1623, he reported, "I did force my selfe to cast & did cast exceding mutch & did I thank god fynd mutch good by it." Ashm. 218, fol. 217.

79. Sloane 3822, fols. 6–20.

80. Here Forman drew the signs engraved on the ring, as appears in figure 3.

81. Sloane 3822, fol. 11. Ashmole, too, has carefully recorded the times of casting. See, for example, Josten, *Elias Ashmole*, 2:546–49.

82. The division of *planetary hours* was based on calculating the time from sunrise to sunset and dividing the number of daylight minutes by twelve. The first and eighth *planetary hours* of each day were associated with the planetary ruler of that day, defining therefore the preferred time frames for astral-magic activity. Skinner, *Terrestrial Astrology*, 240.

83. Sloane 3822, fol. 23.

84. Sloane 3846, fol. 45.

85. Kieckhefer, *Magic in the Middle Ages*, 132–33; Page, "Image-Magic," 77.

86. Josten, *Elias Ashmole*, 4:1523.

87. This included sulfur for Saturn, crocus for Jupiter, piper for Mars, red sanders and crocus for the Sun, costus for Venus, mastick for Mercury, and aloes for the Moon. Ashm. 421, fol. 154v.

88. Zambelli, *Speculum Astronomiae*, 241. On the use of incenses and perfumes in the practice of magic, see also Baker, *Cunning Man's Handbook*, 448–51.

89. On the "exchange value" of sigils, see Kassell, "Economy of Magic," 43–48.

90. Hiebner, *Mysterium Sigillorum*, 159–60. Curiously, Hiebner's primary mentioned source is Theophrastus, a disciple of Aristotle.

91. The first recorded use of a talismanic cure in Napier's medical practice appears on the last day of October 1611, just one month after his mentor's death. Following a short list of herbal medicines, prescribed for Alice Abrys—a thirty-three-year-old woman servant from Stoke Hammond who complained of "a noyse in her hed"—Napier noted advice by the archangel Raphael to "give it a sigill beaten into powder to serve four tymes." Ashm. 200, fol. 183. From this point on, sigils begin to appear in the diaries in erratic proportions. Peaking in April 1612, they temporarily fade in November of that year only to reappear with renewed vigor eighteen months later. In June 1614, the hitherto popular talisman of Venus is replaced by the Jovian sigil, which becomes henceforth the identifying mark of Napier's magical healing.

92. For some typical examples, see Ashm. 233, fols. 26, 33v, 143.

93. Ashm. 421, fol. 171v.

94. Ibid. In one recipe, Napier combined a Jovian sigil with "6 horse leaches" and "cold losenges [troches]." Ashm. 233, fol. 143 (Campren).

95. Hiebner, *Mysterium Sigillorum*, 166.

96. Sloane 3822, fol. 22.

97. Sloane 3846, fol. 102. I could not find, however, any significant correlation between the ruling celestial forces in Napier's astrological charts and his prescription of sigils.

98. Some typical timing instructions given by Napier for the application of sigils are the following: "ap[ply] *die* [in the day of] ♂," Ashm. 416, p. 294 (Mun); "tomorrow about six in the mor[ning]," Ashm. 233, fol. 143 (Wibmer); "to be put on the 17 of Novemb[er]," Ashm. 405, fol. 317 (Robynson); "only the day long," Ashm. 233, fol. 143v (Kamclen); "Wed[nesday] next half an houre before sun set," Ashm. 235, fol. 160 (Whitlerke). For similar directives by Ashmole, see Sloane 3822, fol. 23.

99. For an example of the latter, see Sloane 3822, fol. 11.

100. Sloane 3822, fol. 35 contains "A prayer before the putting on of any sigill," by Sir Thomas Myddelton, Napier's brother-in-law and his apprentice in astral magic.

101. Hiebner, *Mysterium Sigillorum*, 180.

102. Ashm. 421, fol. 171. Gout and dropsy were considered afflictions governed by Saturn, which also ruled the bones (hence apophysis). This suggests a working by antipathy. Most kinds of the falling sickness and wind colic were considered lunar diseases. Eland, *Tutor to Astrologie*, 12–13, 19–20; Gadbury, *Thesaurus Astrologiæ*, 29, 32.

103. Ashm. 213, fols. 14 (Fosters), 61 (Child), 158 (Wyllyson), 166 (Underhill), 167v (Lollyns); Ashm. 412, fol. 173 (Dodsworth); Ashm. 414, fol. 31v (Knight); Ashm. 416, p. 210 (Katlen). This particular directive is often twinned with "appensile" or "sarcenet." See figure 4; and Ashm. 213, fol. 14 (Fosters); Ashm. 235, fols. 159v (Kings), 160 (Whitlerke).

104. Ashm. 213, fol. 166 (Basse); Ashm. 233, fol. 116 (Michels); Ashm. 414, fol. 210 (Katlen).

105. Ashm. 213, fols. 61 (Child), 158 (Wyllyson); Ashm. 233, fols. 143 (Lady Wilmer), 143v (Kamclen), 144v (Gadsene). In some prescriptions, the reading appears more like "seric," which might be an abbreviation of the archaic "sericated" ("clothed in silk"). See, for example, Ashm. 213, fols. 166 (Underhill), 167v (Lollyns); Ashm. 413, fol. 223v (Maple).

106. Ashm. 213, fol. 14 (Fosters); Ashm. 414, fols. 31v (Knight), 183 (Lengland).

107. Spence, *Encyclopaedia of Occultism*, s.v. "St. John's Wort," 345.

108. See, for example, Napier's prescription for a Jovian sigil wrapped in taffeta silk and another "to drinke in *aq[ua] hyper[icum]*." Ashm. 213, fol. 113v.

109. Ashm. 421, fol. 171. See also Napier's sigillated cakes in the example above.

110. Sloane 3846, fol. 46; Hiebner, *Mysterium Sigillorum*, 166. See also Baldwin, "Toads and Plague," 232; Kieckhefer, *Magic in the Middle Ages*, 67; Page, "Image-Magic," 77; Peterson, *Clavis*, 289, 335, 340, 368. In one case, Napier simply prescribed the sigil to be "ap[plied] in silke." Ashm. 230, fol. 5 (Sanders).

111. A note from the summer of 1617 that reads "appens in silke" suggests the former. Ashm. 220, fol. 111v. See also Napier's order for a "S[igil of] ♃ tyd [tied] about his necke." Ashm. 196, fol. 99v (Banbery).

112. Ashm. 405, fol. 236v (Tilcock). See also Ashm. 405, fols. 260v (Canthorne), 317 (Robynson); Ashm. 413, fol. 223v (Maple).

113. Although some astral treatments did have a visible effect. For example, Ashmole once noted that his wife "vomited the first time after she had her sig[il] put on, but brought up wonderful thick phlegm such as was not usual to bring up." Ashm. 421, fol. 102, quoted in Josten, *Elias Ashmole*, 4:1513.

114. Ashm. 416, p. 311. The original consultation notes are at p. 294.

115. Dunte, *Decisiones*, 220, quoted in Clark, *Thinking with Demons*, 485.

116. Perkins, *Salve for a Sicke Man*, 132–33. See also Wear, "Religious Beliefs," 157–58.

117. Thomas, *Religion and the Decline*, 500.

118. Ashm. 414, fol. 15v; Ashm. 196, fol. 103; Ashm. 220, fol. 40v.

119. Augustine, *De Doctrina Christiana*, book 2, 23.36 and 29.45, quoted in Copenhaver, "Scholastic Philosophy," 528. See also Graf, "Augustine and Magic."

120. Zambelli, *Speculum Astronomiae*, 241–51.

121. Weill-Parot, "Astral Magic," 170–77. See also Pingree, "Learned Magic," 41–43. On Thomas's view, see Weill-Parot, "Astral Magic," 175; Kieckhefer, *Magic in the Middle Ages*, 131–33. Some of the difficulty in establishing a clear division between demonic magic and licit astrological images was occasioned by the fact that the Thomist criterion for distinguishing between the two was seen as merely circumstantial.

122. "What was at issue here," observes Keith Thomas, "was not any difference of religious principle; it was a view of the natural world." Thomas, *Religion and the Decline*, 253–79, here 255–56.

123. Clark, *Thinking with Demons*, 472–93, 509–25. Cf. O'Neil, "*Sacerdote ovvero strione*."

124. Men and women like Napier, who provided help through acts of magic, were seen by certain ecclesiastical speakers as a greater threat to divine and terrestrial order than the despised black witches, since their remedies were positively alluring and therefore endangered the many. Clark, *Thinking with Demons*, 457–71.

125. Aston, *England's Iconoclasts*, 1:343–445. See also Eire, *Against the Idols*, 89–94; Dyrness, *Reformed Theology*.

126. Aston, *England's Iconoclasts*, 1:401–8, here 446.

127. Sloane 3822, fols. 65–67v. The text, neither signed nor dated but in Napier's hand, is titled *De sc[ien]do precepto* (A Precept to Be Known) and is part English, part Latin. Its style and wording, as well as the fact that it is interspersed with numerous references to the same church fathers and theologians that Napier employs throughout his theological writings, suggest strongly that it is indeed his original formulation.

128. But note that, as a number of recent commentators have shown, Protestantism was not as iconophobic a religious culture as has sometimes been supposed. See Green, *Print and Protestantism*; Walsham, "Impolitic Pictures"; Watt, *Cheap Print*, 178–216. Furthermore, Walsham identifies an ecclesiastical lowering of the stringency of tone adopted toward idolatry and images in the early Stuart period. Walsham, "Angels and Idols," 154–55. For Napier's sources of inspiration, see also Aston, *England's Iconoclasts*, 1:452–66.

129. Sloane 3822, fol. 65. Prometheus, Ninus, and Epiphanius are the only nonbiblical authorities mentioned in this pamphlet. Ninus was the Chaldean king in the lifetime of Zoroaster, to whom Napier also refers as "governour of the Assyrians." Ashm. 242, fol. 187v. The notion of Haran's image being worshipped by his father (mistakenly identified by Napier

as Nachor instead of Terah) is based on a traditional reading of Wisdom 14:15–16: "For a father afflicted with vntimely mourning, when he hath made an image of his childe soone taken away, now honoured him as a god. . . . Thus in processe of time an vngodly custome growen strong, was kept as a law, and grauen images were worshipped by the commandements of kings."

130. Sloane 3822, fol. 65.

131. Ibid.

132. Ibid. A very similar approach is voiced in Montague, *New Gagg*, 299–319, esp. 299–300.

133. These are based on Leviticus 26:1 and 19:4, Deuteronomy 4:16 and 5:8, Jeremiah 10, Wisdom 14:8, Psalms 97:7, and Baruch 6:47.

134. Sloane 3822, fols. 65r–v.

135. Sloane 3822, fol. 65v.

136. Ibid., based on 2 Kings 7:13–15.

137. Sloane 3822, fol. 65v. Solomon's temple had carvings of "lions, oxen, and cherubim" (1 Kings 7:29).

138. Sloane 3822, fol. 65v.

139. "Render therefore unto Caesar the things which are Caesar's; and unto God the things that are God's." Matthew 22:21. Scriptural examples include the stone set in Bethel by Jacob after his miraculous dream (Genesis 28:18–22), the twelve stones set up in Gilgal and later under an oak tree in Sechem by Joshua (Joshua 4:20; 24:26), and the stone set up as a remembrance of God's help by the prophet Samuel (1 Samuel 7:12).

140. Sloane 3822, fol. 65v. Cf. Napier's argument on the theological legitimacy of astrology in his "Defence of Astrology." Ashm. 242, fols. 190, 192.

141. Sloane 3822, fol. 66v, based on "To whom then will ye liken God? or what similitude will ye set vp vnto him?," Isaiah 40:18, and on 1 Timothy 1:17, 6:16.

142. Sloane 3822, fol. 66v. Cf. Montague, *New Gagg*, 304–7.

CHAPTER 3

1. Britton, *Memoire*, 42–48; Kiessling, *Life of Anthony Wood*, 108; Balme, *Two Antiquaries*; Parry, "Wood, Anthony." On Aubrey's life and work, see Hunter, *John Aubrey*; Poole, *John Aubrey*; Woolf, *Social Circulation*.

2. Wood himself had dined at Ashmole's house in May 1670, while the latter had "shewed him all his rarities, viz. antient coins, medals, pictures, old MSS. &c." Josten, *Elias Ashmole*, 1:172.

3. See, for example, his list of "Nativities noted downe by Dr. Napier in his diarys." Ashm. 423, fols. 60r–v, 62–73.

4. Oxford, Bodleian Library, MS Aubrey (hereafter Aubrey) 6, fol. 10v, cited in Josten, *Elias Ashmole*, 4:1685.

5. Josten, *Elias Ashmole*, 1:244 and 4:1697–98.

6. Oxford, Bodleian Library, MS Tanner 456a, fols. 34–35, cited in Josten, *Elias Ashmole*, 4:1859–61. According to Josten, "The Dreams" could be the lost "Collection of divine Dreams from persons of my acquaintance, worthy of beliefe, 8°," mentioned within "A Catalogue of Books Writen by Mr. Aubrey," in Aubrey 5, fol. 123v. Josten, *Elias Ashmole*, 4:1860n6. It might also refer to materials he initially collected for the chapter on "Dreams" in his own *Miscellanies*.

7. Wood, *Athenae Oxonienses*, 2:103–4; Aubrey, *Miscellanies*, 133–35. Aubrey's report was quoted shortly after, with acknowledgment, in Turner, *Compleat History*, 15. Aubrey also published an abridged account of Napier's angelic communications in his renowned *Brief Lives*. Aubrey, *Brief Lives*, 217–18. This account included further information that did not appear in the *Miscellanies*.

8. For this new insight on Lilly's short narratives, see Raymond, *Milton's Angels*, esp. 106, 126–27.

9. Lilly, *Last of the Astrologers*, 22, 27, 47–48, 93–97. Napier was among the recommenders of John Evans's "Magneticall or Antimoniall Cup." See Evans, *Universall Medicine*, sig. C3.

10. Raymond, *Milton's Angels*, 127.

11. Lilly, *Last of the Astrologers*, 50. On this curious comment of Ashmole's, see note 56 below.

12. The selection of materials surveyed in this section constitutes a representative sample of later discussions of Napier's angelomancy.

13. "That there are *Angels* and *Spirits*, both good and evil," wrote the late seventeenth-century astrologer John Gadbury, "I never yet questioned, nor ever knew any so to do.... But that there are either Angels or Spirits, good or bad, within the compass or power of man's invocations or commands, I seriously protest and I am yet to believe." Gadbury, *Natura Prodigiorum*, 163–64. See also my discussion on angels in the Reformation toward the end of this chapter.

14. Aubrey, *Miscellanies*, 133–34. Cf. Aubrey, *Brief Lives*, 217–18.

15. Davies, "Angels," 312–15; Raymond, introduction to *Conversations with Angels* and *Milton's Angels*; Walsham, "Angels and Idols" and "Invisible Helpers."

16. Napier's "Call" is mentioned under the section for "Visions in a Berill, or Crystall," in Aubrey, *Miscellanies*, 129.

17. Some sheets contain only questions, while others provide answers for queries that appear elsewhere or are missing altogether.

18. John Prideaux (1578–1650), a renowned Calvinist theologian, was Regius Professor of Divinity at Oxford, Rector of Exeter College, and from 1641 the bishop of Worcester. *History of the University of Oxford*, 4:457–58.

19. Ashm. 235, fol. 188v.

20. Ashm. 213, fol. 185. Napier's letters to Prideaux, however, contain no mention of this prophecy. See Ashm. 1730, fols. 188–91v.

21. John Aubrey was taught with Prideaux's logic at Trinity College and used his manuals of moral philosophy, which he later praised.

22. Aubrey, *Miscellanies*, 134. On Sir George Booth, see Kelsey, "Booth, George."

23. Ashm. 235, fol. 187. For a similar prediction, supplied to Mrs. Booth in July 1619, see Ashm. 235, fol. 190v.

24. Aubrey, *Miscellanies*, 133.

25. Cadose, *Hidden Chamber*, 50.

26. It does prefix, nonetheless, many of his assistants' prescriptions. See, for example, Ashm. 409, fols. 122v–26v, 129v; Ashm. 237, fols. 66–67.

27. To the best of my knowledge, no other medical diary from the early modern period reveals a similar angelic intervention in an ongoing medical practice.

28. Napier prefixed the verdicts of other angels with their capital or first couple of letters. Perhaps ℞ is a distortion of "Rp" (the first two consonants of "Raphael"), or a mark designed to distinguish Raphael's answers from the "R n" [Richard Napier] that prefixed Napier's own medical notes.

29. Aubrey, *Miscellanies*, 134–35.

30. Ibid., 135.

31. A further confirmation of Aubrey's reading can be gained from Napier's texts of invocation, one of which includes the following directive: "when the angell appeareth then say as followeth: O thou angle of god ℞ most blessed is he from whome thou comest." Ashm. 244, fol. 131. Throughout 1613, Napier commonly prefixed Raphael's verdicts with an "℞ affir" or "℞ aff," probably an abbreviation for "Raphael affirmeth." Ashm. 199, passim. In a further confirmation of Aubrey's reading, Napier once reported disappointedly that "℞ gave no answ[er] to my consult[ation]." Ashm. 237, fol. 17v (Harys).

32. Some inferred evidence suggests that the questions and answers in Ashm. 237, fol. 192v, are from the autumn of 1614, making them the earliest surviving *Interviews*. Remarks in the medical diaries that are prefixed by the ℞ sign date back to at least 1611. A most extraordinary prophecy by Raphael, foretelling a great plague that will sweep London and the rest of Europe three years hence, was allegedly recorded by Napier in January 1606, five years prior to the first occurrence of angels in his diaries. This one-page account of an apocalyptic message is unquestionably in Napier's hand but might have been copied by him from some external source. Sloane 3822, fol. 24. Cf. Uriel's laconic prophecy to Dee, of a "great misserie, to the heavens, to the earth and to all liuing Creatures," quoted in Whitby, *John Dee*, 2:302.

33. Dunton, introduction to *Dunton's Ghost*.

34. Madden, *Memoirs*, 1:233.

35. Lysons, *Environs of London*, 1:303.

36. Lysons and Lysons, *Magna Britannia*, 1:596–67.

37. "Magna Britannia." In 1821, the *Recreative Magazine*, a journal of "eccentricities of literature and life," quoted the *Magna Britannia* on Napier within a discussion on the devotion of priests. "The Rosary." See also *Lambeth and the Vatican*, 2:117–18.

38. Bliss, "Additions," 389. On Philip Bliss, see Campbell, *Retrospective Review*, 7–8. Cf. the 1834 account of the English philosopher and journalist William Godwin, who described Dee's history as "strikingly illustrative of the credulity and superstitious faith of the time in which he lived," Godwin, *Lives*, 390.

39. Davies, *Grimoires*; Harris, *Folklore*. On the renewed interest in spiritualism and occultism in Victorian England, see, for example, Asperm, *Arguing with Angels*, 44–46.

40. Scott, *Letters of Daemonology*. For Scott's references to Napier, see Scott, *Secret History*, 2:232. Both accounts were based on Lilly's autobiography.

41. Horst, *Zauber Bibliothek*.

42. [Moir], "Demonology and Witchcraft," 14.

43. Working mainly as an attorney, Moir devoted considerable time to his literary pursuits and became a regular contributor to *Blackwood's Magazine*. He published, under his own name, a full reprint of the same article: Moir, *Magic and Witchcraft*.

44. [Moir], "Demonology and Witchcraft," 14. The complaint was directed specifically toward Napier; adjacent examples include "the sympathetic nostrums of Valentine Greatrakes and Sir Kenelm Digby" and Aubrey's report on the case of Arise Evans, who was allegedly cured by rubbing his nose against the hand of King Charles II.

45. [Moir], "Demonology and Witchcraft," 14. Some publications that quoted from or reprinted an abridged part of the article in *Foreign Quarterly Review* are *Mirror of Literature, Amusement, and Instruction* 16 (1830): 312; *Polar Star of Entertainment and Popular Science* 5 (1830): 87; *Museum or Foreign Literature and Science* 17 (1830): 230; *Phrenological Journal and Miscellany* 6 (1830): 516; John G. Dalyell, *The Darker Superstitions of Scotland Illustrated from History* (Edinburgh: Waugh and Innes, 1834), 529.

46. For a rather anachronistically sympathetic report, see Smith, *Familiar Astrologer*, 23–25. Later mentions of Napier's angelic conversations include Mark Napier, "Memoirs of John Napier of Merchiston, his Lineage, Life, and Times," *Quarterly Review* 52 (1834): 458 and *Memoirs*, 240–41; Alexandre J. Brierre de Boismont, *Hallucinations: Or, the Rational History of Apparitions* (Philadelphia: Lindsay and Blakiston, 1853), 297; Oliver W. Holmes, *Medical Essays, 1842–1882* (1861; repr., Boston: Houghton Mifflin Company, 1911), 354; Richard Chambers, ed., *The Book of Days: A Miscellany of Popular Antiquities* (London: W. & R. Chambers, 1863), 1:458; Eliakim Littell, *Littell's Living Age* (Boston: Littell, Son, 1865), 243; James Hingston, *Gospel of the Stars* (New York: Continental, 1899), 113; Charles E. Mallet, *A History of the University of Oxford* (1924; repr., London: Methuen, 1968), 1:378.

47. Kate Bennett suggests that Aubrey was a sort of "hint-keeper," who assisted the passing of ingenious ideas to future generations. Bennett, "John Aubrey."

48. For a recent history of the passage of occult ideas and practices in England through the eighteenth and nineteenth centuries, see Monod, *Solomon's Secret Arts*. See also Barry, *Raising Spirits*. Barry tells the story of Thomas Perks, a seventeenth-century gunsmith from Mangotsfield whose conjurations of spirits reportedly led to an early death, and traces the story's transmission through the eighteenth and nineteenth centuries.

49. Raymond, introduction to *Conversations with Angels*, 7.

50. One of the best indicators of Napier's sincere belief in the reality of his spiritual conversations is the fact that he sometimes addressed Raphael with questions concerning himself. See, for example, Ashm. 222, fol. 194v; Ashm. 235, fols. 190v–91v, 192v; Ashm. 414, fol. 188.

51. Raymond, "Tongues of Angels," 271. See also Marshall and Walsham, "Migrations of Angels"; Davies, "Angels"; Principe, *Aspiring Adept*, esp. 188–208. Some of these later seventeenth-century practitioners, who are mentioned in Lilly's autobiography and Aubrey's *Miscellanies*, I have named in the course of this chapter. Another such figure is the astrologer and alchemist John Pordage (1607–1681), whose contacts with angels are discussed in Raymond, *Milton's Angels*, 125–61.

52. Lily, *Last of the Astrologers*, 148–49.

53. Ashm. 1488, part 2, fol. 21v; Ashm. 415, fol. 143. Cf. Dee, *Diaries*, 303n3. Napier resolved in 1602 an astrological question for Dee's son, Arthur, who inquired "whether he shall obtayne his wifes dowery," and found "great grief & discontentment for wants & grief touching his father." Ashm. 221, fol. 51v. For Napier's other engagements with Arthur, who for a short period acted as a scryer for his father, see Ashm. 221, fol. 56; Ashm. 1501, part 4, fols. 5r–v, 6v.

54. See note 9 above; and Aubrey, *Miscellanies*, esp. 128–32.

55. Ashm. 200, fol. 54 (Sir Berkeley).

56. Ashm. 200, fol. 60v (Michels). For other verdicts by Michael, see Ashm. 200, fols. 61v (Caucot), 62 (Moore), 63v (Faldoe, Richardson, and Rooles), 82 (Jackman), 112v (Wels). Michael sometimes gave short medical recipes, which might clarify Ashmole's remark in Lilly's autobiography, quoted above. See, for example, Ashm. 409, fol. 88v (Crips).

57. For example, seven of the eight fully documented cases for April 19, 1611, include such prophetic judgments.

58. Ashm. 200, fol. 62 (Walton).

59. For this multitude of angels, see, for example, Ashm. 200, fols. 114v (Brete), 124v (Fosket), 125 (Markam), 142v (Hall and Parret), 143 (Lame, Richardson, and Brigs), 143v (Write). For some examples of multiple angelic verdicts for the same case, see Ashm. 200, fols. 103v (Dungles), 107v (Cordell), 164 (Morten). Asariel (abbreviated as As or Asar) is the only one of Napier's angels who has no scriptural origins and is not listed among the archangels, and his identification is therefore not entirely clear. It is possible, for example, that Asar represents a rendering of Azarias, a name assumed by the archangel Raphael in the book of Tobit. Napier's notes seem to contain other marks of angels, which I found unintelligible, including W, An, and +.

60. A rather late verdict ascribed to the archangel Michael appears on November 4, 1612, and an even more exceptional one as late as May 1620. Ashm. 409, fol. 197v (West); Ashm. 414, fol. 59 (Gyls). All other verdicts from mid-1612 on are Raphael's.

61. The proportion of angelic assistance, out of the total of daily medical cases in this period, varies from about 30 percent to 70 percent. At specific early dates, such as June 10, 1611, an angel was consulted in almost every case.

62. After March 1612, angelic judgments appear as little as once a month, although one can find relatively dense clusters alongside months of silence. Many of the notes are not prefixed by an angel's capital but keep the same content and form. After 1616 the frequency of angelic notes in the regular diaries declines even further. Earlier that year, Napier reported being called "a conjurer" by a certain "Goody Kent," who met him and his assistant Robert Wallys (or Wallis) on the street. This might have been one of the reasons that led him to conceal his angelic contacts. Ashm. 408, fol. 28. However, some rare verdicts by Raphael appear

as late as 1618. See, for example, Ashm. 201, fols. 43v (Sir Andrewes), 45 (Goetobed), 59v (Simpson).

63. Ashm. 235, fols. 118v–65v, 187–91v; Ashm. 213, fols. 185r–v. I have looked at Napier's medical notes from these particular dates simply because their correlating queries survive in the *Interviews*.

64. The book of Tobit, chaps. 6 and 8. On Raphael's role as a "celestial physician" in Milton's *Paradise Lost*, see Mohamed, *Anteroom of Divinity*, 115–40. According to rabbinical tradition, Raphael was appointed the angel of healing, and as such gave a medical book to Noah after the flood, soothed Abraham's pains after circumcision, and tended to Jacob's thigh after his wrestling with the archangel. See Ginzberg, *Legends*, 1:173–74, 1:385, 4:360. See also Marshall and Walsham, "Migrations of Angels," 11; Keck, *Angels and Angelology*, 170–71.

65. Ashm. 414, fol. 134v (Horwood).

66. Ashm. 416, p. 210 (Katlen). Napier remarked that at times he would come to his senses, "speake well & know his old acquaintance[s]."

67. For Napier's notes on those aspects of the planets that "signify health" and "death," see Ashm. 174, p. 428. Cf. Saunders, *Astrological Judgment*, part 1, 39–66. For astrologically based prognoses given by Forman in medical cases, see, for example, Ashm. 234, fols. 3v (Starkey), 5v (Clark), 6v (Badgly), 14v (Parrsly), 21 (Prat). Ashmole likewise checked astrologically whether his wife "shall recover this sickness." Josten, *Elias Ashmole*, 2:586.

68. Ashm. 414, fol. 139 (Jenyson). This verdict is not prefixed by an angel's capital, but it resembles, in content and form, other angelic verdicts.

69. Ashm. 414, fol. 136v (Dacres).

70. Ashm. 235, fol. 190.

71. Ibid. "This woman had an apple given her by a suspecten woman for a witch," Napier noted in his diary, "& after she tooke it she presently gone mopish, mad & foolish & senceles [senseless]." Ashm. 235, fol. 123v.

72. Ashm. 235, fol. 189v.

73. Ashm. 223, fol. 195v. Cf. "℞ he is past all helpe." Ashm. 199, fol. 20v (Quea).

74. Ashm. 223, fol. 195v. For the corresponding medical note, see Ashm. 416, p. 210 (Katlen).

75. Ashm. 414, fol. 135 (Spencer). Napier's initial note read "will live to Easter day," but he later crossed out the last three words and corrected his verdict.

76. Ashm. 414, fol. 221.

77. Ashm. 414, fol. 188.

78. Ashm. 414, fol. 134v (Hollys).

79. Ashm. 414, fol. 220.

80. According to the Hanslope church records, a man by the name of Thomas Hollys did indeed die there in 1620. *Hanslope Church Register for 1571–1732*, 73.

81. Ashm. 235, fol. 159v (Kings).

82. Ashm. 235, fol. 189.

83. Ashm. 235, fol. 160 (Whitlerke).

84. Ashm. 235, fol. 187v.

85. Ashm. 199, fol. 106v (Andrewes).

86. For some other examples, see Ashm. 199, fols. 13v (Harrys), 106v (Foster), 156v (Shepheard), 165v (Ously); Ashm. 408, fols. 42v (Richard Napier), 101v (Gilpyn).

87. Ashm. 414, fol. 188.

88. Ashm. 223, fol. 195v.

89. Ibid.

90. Ashm. 235, fol. 186v. The cure apparently worked as Napier later noted that "she was mutch the better."

91. Ashm. 222, fol. 194v. No answer is recorded for this query, as for some of the others that follow.

92. Ashm. 213, fol. 185. Such long, detailed prescriptions were more often impersonal. See, for example, Ashm. 200, fol. 242 ("for palsy"); Ashm. 220, fols. 189v, 190v ("for the stone"); Ashm. 402, fol. 186v; Ashm. 408, fol. 180v.

93. Ashm. 235, fol. 19v (Panton). Since Napier makes her the sister of "Mrs Booth," she might be identified as Alice Panton: the daughter of Sir William and Lady Elizabeth Booth and the great aunt of Sir George Booth, whose birth was prophesied by Napier. Burke and Burke, *History of the Baronetcies*, 73.

94. Ashm. 235, fol. 192v. There are clues to suggest that Napier carried out some kind of follow-up on the archangels' advice, in a somewhat similar way to that employed on his own medical instructions. For instance, besides Raphael's instruction, in the case of a young woman named Sara, to "let her blood in the brow," he later recorded that "she bled well in the brow." Ashm. 408, fol. 101v (Gilpyn). See my discussion of Napier's medical follow-up in the chapter on astrological medicine.

95. Ashm. 235, fol. 188. See also Ashm. 235, fols. 187r–v. Elizabeth Marsh, the wife of Thomas Marsh, was the daughter of Napier's sister Katherine. Cokayne, *Some Notice*, 12. For further queries concerning her husband and son, see Ashm. 414, fol. 186. About the disease known as "the mother," see note 8 in the introduction.

96. Ashm. 235, fol. 188v.

97. Ashm. 223, fol. 195v.

98. Ashm. 235, fol. 189v. Sir Thomas Temple (1567–1637), 1st Baronet of Stowe, Buckinghamshire, was an English landowner and member of Parliament. His eldest son, Sir Peter Temple, married Ann, the daughter of Sir Arthur Throgmorton, an English courtier and politician who was Napier's neighbor and patient. Noble, *Lives*, 2:266–67.

99. Ashm. 235, fol. 189.

100. Ashm. 222, fol. 195. Cf. Ashm. 200, fol. 196. "Mr Latch" was the grandfather of Elizabeth Jennings, the girl whose suspected bewitchment opened the introduction. See also his letter to Napier about a pomander given to "little Francke." Ashm. 1458, fol. 129.

101. Ashm. 223, fol. 195v.

102. On the proportion of mental afflictions in Napier's clinic, see MacDonald, *Mystical Bedlam*, 31–32. The medical problem is not always stated in the *Interviews*, making the actual proportion potentially even higher.

103. Ashm. 235, fol. 191.

104. Ibid. Suspected bewitchment constitutes about 12 percent of the surveyed medical queries.

105. Ashm. 222, fol. 194.

106. Ashm. 223, fol. 194v.

107. Ibid.

108. Ashm. 235, fol. 160 (Stuchbury). For the correlating query in the *Interviews*, see Ashm. 235, fol. 187v.

109. Ashm. 235, fol. 187v.

110. Ashm. 200, fol. 240v.

111. Ashm. 235, fol. 187v.

112. Ashm. 413, fol. 106v (Myddleton).

113. For other queries concerning pregnancy, see, for example, Ashm. 235, fols. 190 (Henly), 191 (Tyringham and Temple), 191v (Andrews).

114. Ashm. 235, fol. 188. Cf. her medical note on Ashm. 235, fol. 158 (Ashwell).

115. Ashm. 414, fol. 185v.

116. On Forman's considerations in accepting new patients, see Kassell, *Medicine and Magic*, 148–52.

117. Ashm. 235, fol. 192.

118. Ibid.

119. Ashm. 222, fol. 194.

120. Ashm. 222, fol. 194v.

121. Ashm. 222, fol. 195.

122. Ashm. 223, fol. 192v. William Page, Doctor of Divinity and Fellow of All Souls College, corresponded with Napier on theological matters in the late 1620s.

123. Ashm. 223, fol. 195.

124. Although there are some exceptions. See, for example, Ashm. 414, fol. 135 (Spencer).

125. Kieckhefer, *Forbidden Rites*, 100. The surviving *Interviews* completely lack inquiries about clients' lost property, yet Napier himself sometimes approached the archangel when he could not trace a lost book. See, for example, Ashm. 222, fol. 197.

126. On the different methods of early modern divination and fortune-telling, see esp. Thomas, *Religion and the Decline*, 237–44.

127. Ibid., 242.

128. Forman's clients, on the other hand, posed such queries only very rarely. See, for example, Ashm. 219, fols. 66, 105.

129. Ashm. 235, fol. 187.

130. One outstanding example is Ashm. 235, fol. 187. This type of repetitiveness is also present in Ashmole's diaries: see, for example, the queries regarding Alexander Blagrave's marriage in Ashm. 374, fols. 82, 83, quoted in Josten, *Elias Ashmole*, 2:557.

131. It is estimated that up to the late seventeenth century the general population of England outlived the peerage. Riley, *Rising Life Expectancy*, 139–40. See also Wrigley and Schofield, *Population*.

132. Ashm. 235, fol. 188. Cf. Ashm. 235, fols. 187r–v.

133. Ashm. 235, fol. 191v.

134. Ashm. 235, fol. 191.

135. Thomas, *Religion and the Decline*, 241. For similar questions resolved astrologically in Forman's consulting room, see, for example, Ashm. 236, fols. 167 (Holmes), 172 (Crowder); Ashm. 411, fol. 52 (Borman). See also Forman's method to "knowe which shal liue longeste of the man and his wife by the letters of their names," using a wheel diagram, in Ashm. 354, fol. 183. A different method, based on the number of letters in the couple's names, is described in Ashm. 174, pp. 375–79.

136. On early modern remarriage rates, see Wrigley and Schofield, *Population*, 190, 258–59, 359–60, 426–27. The authors suggest that as many as 30 percent of all those marrying in the mid-sixteenth century were widows or widowers.

137. Ashm. 235, fol. 188.

138. Ashm. 235, fol. 187.

139. Ibid.

140. Ashm. 235, fol. 191.

141. For marital queries solved astrologically by Forman, see, for example, Ashm. 234, fols. 6 (Davis), 9v (Kitchen), 25 (Heaton), 39 (Moody), 82v (Breed), 83 (Bassano). For similar questions posed to Ashmole, see, for example, Ashm. 374, fols. 75v, 82, 83, quoted in Josten, *Elias Ashmole*, 2:554, 557.

142. Ashm. 235, fol. 188.

143. Ashm. 235, fol. 190.

144. Ashm. 414, fol. 187.

145. Ashm. 414, fol. 187v.

146. Ashm. 222, fol. 197.

147. Ibid.

148. Ashm. 235, fol. 191v.

149. Ashm. 414, fol. 221.

150. Ashm. 414, fol. 186.

151. Ashm. 235, fol. 187. Sir Robert Napier (1603–1661), 2nd Baronet of Luton Hoo, knighted in 1623 and a member of Parliament from 1625, was the eldest son and namesake of Napier's elder brother. Sir Richard was Sir Robert's younger brother and Napier's future successor and heir.

152. The same prophecy and subsequent correction recur almost identically in Ashm. 235, fol. 191.

153. For other instances in which Napier recorded an angel's change of mind, see, for example, Ashm. 200, fol. 95v (Searell); Ashm. 220, fol. 187v (Earl of Kent); Ashm. 237, fol. 65 (Moore).

154. Ashm. 235, fol. 191. A similar query recurs on the same page: "Wheather Sir Jane shall out live his brother John & [whether] his brother John [shall] die without children." ℞: "he shall & shall have no child."

155. Ashm. 235, fol. 188.

156. Ashm. 240, fol. 135.

157. Ashm. 414, fol. 159v (Evington). Napier later annotated the record to indicate that Evington "dyd sell his lands."

158. Ashm. 414, fol. 220v.

159. Ashm. 414, fol. 137v (Evington).

160. Ashm. 414, fol. 187.

161. Ibid.

162. Ashm. 414, fol. 138 (Evington). Sir James eventually sold his house to William Trollop in April 1621. About a month earlier, probably in an expression of remorse, he wrote a letter to his uncle renouncing "the sporte of running horses" and other "idle expenses." Ashm. 240, fol. 126. According to Michael MacDonald, Sir James Evington was also the authentic addressee of Napier's letter on usury from 1622. MacDonald, "Defence of Usury," 356.

163. See, for example, Ashm. 223, fol. 192v (Mounsen); Ashm. 235, fols. 188 and 189 (Mydleton), 192 (Rush); Ashm. 414, fol. 220v (Andrews).

164. Ashm. 235, fol. 191v. Cf. his queries concerning the works of Raymond Lull, John of Rupescissa, and Constantine Albinius, in Ashm. 235, fols. 186v, 188v.

165. Ashm. 235, fol. 191v. Napier repeats the same question, in two slightly different versions that are no less enigmatic, in Ashm. 235, fol. 191.

166. Ashm. 374, fol. 52, quoted in Josten, *Elias Ashmole*, 2:536.

167. Ashm. 223, fol. 195v.

168. Ashm. 223, fol. 194v.

169. See, for example, Ashm. 213, fol. 185v and Ashm. 235, fol. 188 (the Digbies); Ashm. 235, fol. 189v (the Dewers).

170. For some of Napier's astrological and medical correspondence, see, for example, Ashm. 174, p. 467; Ashm. 240, fol. 78; Ashm. 1458, fols. 119, 129, 155; Ashm. 1501, part 4, fols. 5–6v.

171. Aubrey, *Miscellanies*, 129.

172. Boudet, *Entre science et nigromance*, esp. 539–56; Klaassen, "Medieval Ritual Magic" and *Transformations of Magic*; Láng, "Angels Around the Crystal"; Page, *Magic in the Cloister*, esp. 131–40. Klaassen specifically stresses the continuities between medieval and Renaissance traditions of ritual magic, especially in regard to technique.

173. The practices of Kabbala include the calling of angels and the ten *sephiroth*, each of which represents a distinct emanation or quality of God. Brach, "Magic."

174. Gordon, "Renaissance Angel."

175. Davies, "Angels," 309–10. See also Clark, *Thinking with Demons*, 214–32.

176. Roos, "Magic Coins," 278.

177. Ashm. 1790, fol. 39.

178. The use of a capital N as a placeholder for the name of the magus (or sometimes for the names of angels) recurs in several other magical templates. See, for example, Add. 36674, fol. 145, and Sloane 3846, fol. 69v.

179. From Napier's text of invocation and instructions for summoning angels (apparently a template), which he copied from "an old booke," Ashm. 409, fol. 221v. The entire text is crossed out, perhaps in light of its clear use of Catholic formulae.

180. A recent exploration of magical traditions and their transformations through the prism of magical texts is Klaassen, *Transformations of Magic*. Other excellent surveys of medieval magic texts are Klaassen, "Medieval Ritual Magic"; Boudet, *Entre science et nigromance*.

181. Davies, *Grimoires*, 63–64; Kassell, *Medicine and Magic*, 215. See also Ashm. 407, fol. 150; and Ashmole's remark in Lilly, *Last of the Astrologers*, 31.

182. Conducting a long session of prayer as part of a medical encounter was a distinctive mark of Napier's practice, earning him the pious status of one whose "knees were horny with frequent praying." Aubrey, *Miscellanies*, 135.

183. Ashmole's copy of another invocation text, taken out of the medieval grimoire *Sepher Raziel*, is titled "The invocation of Oberion concerning Physic &c," but records the substitution of the word "Oberion" with "Raphael." Beside this, Ashmole noted, "so mended with Dr Napiers hand." Sloane 3846, fol. 102v. Cf. the full text, with the same amendment, inside Gerence James's copy of *Sepher Raziel*, Sloane 3826, fols. 98–99. See also Napier's prayers of invocation in Ashm. 1790, fols. 112r–v, 113v, 115. For a useful (if somewhat arbitrary) survey of the operation of invocation in the English scene up to the nineteenth century, see Baker, *Cunning Man's Handbook*, 462–76.

184. See also Ashmole's directions for repeating the same devotional prayers. Sloane 3846, fol. 67.

185. The quote is from Láng, "Angels Around the Crystal," 7. See also Napier's direct appeals to God or to Christ in Ashm. 237, fol. 190v.

186. Thomas, *Religion and the Decline*, 268; Walsham, "Invisible Helpers," 80–86.

187. Prideaux, *Patronage of Angels*, 23, quoted in Walsham, "Catholic Reformation," 281. On Casaubon's campaign against Dee's conversations, see Harkness, *Conversations*, 222–25.

188. Hall, *Cases of Conscience*, 175.

189. Walsham, "Invisible Helpers," 108–20.

190. Thomas, *Religion and the Decline*, 267–72.

191. On medieval and early modern theurgy, see Fanger, "Introduction"; Louth, "Pagan Theurgy"; Véronèse, "Magic, Theurgy, and Spirituality."

192. Marshall and Walsham, "Migrations of Angels," 50; Véronèse, "Magic, Theurgy, and Spirituality," 50–52. See also Klaassen, *Transformations of Magic*, 91, 141, regarding these requirements in the practice of ritual magic in general.

193. Ashm. 1790, fol. 39.

194. Lilly, *Last of the Astrologers*, 96.

195. On the appropriate conduct for a practitioner of angelic magic, see also Klaassen, "Ritual Invocation," 343.

196. Early modern English magi saw, nonetheless, the requirement of virginity for the scryer as a Catholic element that should be stripped off (see below). On virgin scryers, see Fanger, "Virgin Territory."

197. Anderson, *Book of Examinations*, 104–5, quoted in Thomas, *Religion and the Decline*, 268.

198. Reynes, *Commonplace Book*, 169ff., quoted in Kieckhefer, *Forbidden Rites*, 97.

199. Klaassen, "Ritual Invocation."

200. Lilly, *Last of the Astrologers*, 95. A copy of Sarah Skilhorn's "call" in Ashmole's hand, titled "To call a good Angell into A christall stone," appears in Sloane 3846, fol. 94.

201. Ashm. 1491, fols. 884, 1127, quoted in Kassell, *Medicine and Magic*, 216. On Forman calling angels and spirits, see Kassell, 215–21.

202. One exception is in his reference to a commentary on Hosea, where he notes that biblical prophets "had their seer, *viz.* young youths who were to behold those visions." Aubrey, *Miscellanies*, 128. This, however, refers to ancient rather than contemporary invocations.

203. Aubrey, *Miscellanies*, 129–31.

204. Ibid., 130.

205. In a document from 1651 (now kept in the Science Museum, London), Nicholas Culpeper describes how John Dee's crystal was bequeathed to his son Arthur Dee, and how he himself later came to be the stone's owner.

206. On the transmission of magical objects among English practitioners of the seventeenth century as means of transferring obscure knowledge, see also Kassell, "Economy

of Magic"; Fanger, *Conjuring Spirits*; Whitby, *John Dee*; Stephenson, "From Marvelous Antidote."

207. Clucas, "False Illuding Spirits," 152; Whitby, *John Dee*, 1:76.

208. Kieckhefer, *Forbidden Rites*, 107. For prayers "to consecrate the glasse," see Add. 36674, fol. 86. Lilly reports of some spirits that were displeased by the lack of suffumigation at the time of their invocation and of a scribe named John a Windor, who would at times "fumigate with Contraries" to vex the spirits. Lilly, *Last of the Astrologers*, 22, 93.

209. Lilly, *Last of the Astrologers*, 96 (in a note by Ashmole).

210. Davies, "Angels," 305-6.

211. Ashm. 421, fol. 165.

212. On the thirteenth-century dispute over angels' corporeality, see Keck, *Angels and Angelology*, esp. 93-114. On angelic bodies and their means of communication with humans, see Raymond, *Milton's Angels*, 69-70, 284-91, 315-24. In the following chapter I briefly discuss Napier's letter on the lawfulness of certain Catholic assemblies, wherein he rejects the decree made by the Fourth Lateran Council that implied that angels were not corporeal in any way.

213. Such views were specifically voiced by Thomas. Aquinas, *Summa Theologiae*, 9:31-43, quoted in Raymond, "With the Tongues," 264-67.

214. Lilly, *Last of the Astrologers*, 83, 198. Aubrey's account, quoted above, regarding a beryl stone in which one could view "either the Receipt in Writing, or else the Herb," suggests one way in which complex prescriptions and treatments could have been communicated to Napier.

215. Keck, *Angels and Angelology*.

216. Partee, *Theology of John Calvin*, 71-72; Meier-Oeser, "Medieval, Renaissance, and Reformation Angels," 198-99.

217. On some seventeenth-century discussions of angels, see Coudert, "Henry More"; Hutton, "Of Physics"; MacDonald Ross, "Occultism and Philosophy." One excellent contemporary example is John Salkeld's 1613 treatise on "the Natvre, Essence, Place, Power, Science, Will, Apparitions, Grace, Sinne, and all other Proprieties of Angels." Salkeld, *Treatise of Angels*. On the varied Protestant attitudes toward angels, see Raymond, *Milton's Angels*, part 1; and the discussion below.

218. The background for most of Napier's questions was specific contemporary debates over the nature of angels. See Raymond, *Milton's Angels*, 65-88.

219. Ashm. 237, fol. 192v.

220. Ibid.

221. Ashm. 200, fol. 240v.

222. Ashm. 237, fol. 192v.

223. Ashm. 200, fol. 240v. The devil, explained John Salkeld, is "euery where present, suddenly flying and passing through the whole world. Almost, though not altogether, with the same breuitie, that our cognition is now here & in a moment is transported to the furthest corner of the world." Salkeld, *Treatise of Angels*, 69-70.

224. 1 John 2:18, 2:22, 4:3; 2 John 1:7.

225. Hill, *Antichrist*, 178-81. On the Antichrist and the Jews, see also Gow, *Red Jews*; McGinn, *Antichrist*. For another conjecture that rather made the Antichrist a Turk, see Montague, *New Gagg*, 75; Aylmer, *Harborovve*, sig. Q. See also Hill, *Antichrist*, 181-82.

226. On the pope as Antichrist, see Hill, *Antichrist*, 1-40; Milton, *Catholic and Reformed*, 93-110; Firth, *Apocalyptic Tradition*; Christianson, *Reformers and Babylon*. John Whitgift himself proved the pope to be the Antichrist in his D.D. thesis. Strype, *John Whitgift*, 1:15. In fact, the majority of English scholars in the first half of the seventeenth century, including the astrologers Nicholas Culpeper, William Lilly, and John Booker, believed the pope was the Antichrist.

227. Napier himself reported a story on a "wandering Jew" in his diary for 1603. Ashm. 207, fol. 59v. On yet another legend, see Napier's question to Raphael "wheather Spinx [the

Sphinx?] can take away mens child out of the cradle & m[ake] them to goe invisible & to goe through doors." Ashm. 223, fol. 192v.

228. Ashm. 237, fol. 192v. Distinguishing perhaps between the English attitudes toward the pope (with which he identified) and the historical figure of the Antichrist, Napier condemns the Church of Rome in one of his theological expositions for "the deniall of the true & only headship of Christ by placing antich[rist] in his roome." Ashm. 1473, fol. 452.

229. Ashm. 414, fol. 185v.

230. Ibid.

231. Ibid. The content and wording of Napier's questions reveal his fundamental belief in the medieval, Roman narrative.

232. Ashm. 237, fol. 188v.

233. Ashm. 200, fol. 240v.

234. Ashm. 223, fol. 192v. See also Napier's undated letter, attacking the "popish" belief in transubstantiation in the Holy Eucharist, in Ashm. 1730, fols. 184r-v.

235. Ashm. 237, fol. 192v.

236. Ashm. 200, fol. 240v.

237. Ashm. 222, fol. 1.

238. Ashm. 223, fol. 192v.

239. Ashm. 414, fol. 185v; based on 1 Kings 13:18.

240. Ashm. 235, fol. 189.

241. Ashm. 235, fol. 191v.

242. Ashm. 414, fol. 220. Raphael's answer was later quoted by Ashmole as a rule for making sigils. See Ashm. 421, fol. 171v.

243. Ashm. 235, fol. 190v. On December 3, 1614, Napier wrote, "℞ gave me 2 Ru[n]dlets." Ashm. 237, fol. 130. It is not clear what this "Rundle of Raphael" was. Perhaps it was some kind of lamin made according to the archangel's directions or bearing his signs.

244. Ashm. 235, fol. 191v.

245. Ashm. 235, fol. 192.

246. Ibid. *Orthopnoea* means "shortness of breath."

247. Ashm. 235, fol. 188. A "quartan fever" was the term used for a malarial fever with paroxysms recurring every three days.

248. Biringuccio, *Pirotechnia*, 35, quoted in Clulee, *Natural Philosophy*, 197.

249. Casaubon, *True and Faithful*, sig. E7.

250. Ashm. 222, fol. 195. Napier probably refers to Forman's copy of "Of Cako," which, according to Lauren Kassell, was "an anonymized and altered version of Alexander von Suchten's 'Second treatise on antimony.'" Kassell, *Medicine and Magic*, 173–89.

251. Ashm. 223, fol. 193v. *Aurum potabile* is potable gold: a therapeutic preparation made of gold that was dissolved in oil and alcohol.

252. Ashm. 414, fol. 47. "My Lord Chancellor" seems to be no other than Sir Francis Bacon, who was lord chancellor of England from 1617 to 1621. "The Earl of Exeter's son" is most likely William Cecil, the eldest son of Thomas Cecil, 1st Earl of Exeter and Bacon's uncle by marriage. According to Eric Ash, Bacon "frequently expressed his respect for those who had personal experience in exploring nature or possessed practical skills and knowledge." Ash, *Power*, 211.

253. The 1604 Act Against Conjuration and Witchcraft sentenced to death all those who shall "use, practice, or exercise any invocation, or conjuration, of any evil and wicked spirit, or shall consult, covenant with, entertain, employ, feed, or reward any evil and wicked spirit, to or for any intent or purpose." Tomlins and Raithby, *Statutes at Large*, 4:599.

254. Heyd, *Be Sober*.

255. Mohamed, *Anteroom of Divinity*; Walsham, "Invisible Helpers," 77–86; Marshall and Walsham, "Migrations of Angels," 33–40.

256. For Luther's attitudes toward angels, see Soergel, "Luther on the Angels."

257. Walsham, "Invisible Helpers"; Marshall and Walsham, "Migrations of Angels"; Raymond, *Milton's Angels*.

258. Marshall, "Angels Around the Deathbed"; Mayr-Harting, *Perceptions*; Sangha, *Angels and Belief*; Walsham, *Catholic Reformation*, 207-34.

259. Iribarren and Lenz, introduction to *Angels*, 2-3.

260. Grant, *Planets*. Napier himself iterates the notions of cosmic ordering and principal and secondary causes in his "Defence of Astrology," stating that God "hath ordained that there should bee first & second causes, & that the world in his governement should resemble a clocke in his motion, & god as the principall wheele drawing on the second, & they the rest, working as the first agent, & principall cause, using the heavens & the starres therein as a second cause in the moving, altering, & disposing of thinges beneath." Ashm. 240, fol. 190.

261. Hallacker, "On Angelic Bodies," 210. Hallacker frames her discussion within the contemporary debate over body and mind, which is less relevant for my purposes.

262. The idea that angels had precedence in the divine hierarchy but did not partake in man's intimate relation with God preceded Protestantism and sometimes spanned the Protestant-Catholic divide. See Bruce Gordon's discussion of the fifteenth-century German philosopher and theologian Nicholas of Cusa. Gordon, "Renaissance Angel," 43-47.

263. Ashm. 1148, p. 204. Cf. Sibbes, *Light from Heaven*, 106. See also Raymond, *Milton's Angels*, 83-84; Harvey, "Role of Angels," esp. 3-4. According to Raymond, Harvey "detects a shift in the perceived relationship between angels and men, from angelic superiority, through equality, to a claim of human superiority through Christ's Atonement." Raymond, introduction to *Conversations with Angels*, 8.

264. Torrance Kirby, *Richard Hooker*, esp. 29-43. Hooker's discussion of the divine ordering (*lex divinitatis*) is given as background to his presentation of the church ecclesiastical hierarchy as an earthly parallel to the celestial one.

265. Lilly, *Last of the Astrologers*, 47.

CHAPTER 4

1. "It was my happ (your coming vnto mee in the waye of Phisike, to cure a bad payne of lunges)," wrote Napier to one of his correspondents, "that vpon some bye talke & discourse at table wee cam[e] into many questions." Napier to an unknown recipient, date unknown, Great Linford, Ashm. 1730, fol. 206.

2. "I went to Mr Byrd's," Napier recorded in his diary for June 8, 1602, "& after a good supper we had [a] talke of the crosse of Jesus christ & greate Disputation, wheather it ough[t] to be used in baptisme I [aye] or noe. [T]hey Denyed it, I [assu]med it, they held it as a popish & a wicked ceremony." Ashm. 221, fol. 80v. See also Ashm. 220, fol. 40v; and below in this chapter.

3. Lilly, *Last of the Astrologers*, 49. As noted earlier, Napier himself admitted his reluctance to preach. Many "godly" Christians were appalled by such nonpreaching clergymen. For some contemporary sources, see Cressy and Ferrell, *Religion and Society*, 98, 109-12, 114-19.

4. One rare exception, concerning the use of the cross in baptism, is found in note 2 above.

5. All of Napier's dated theological writings are from the 1620s. In fact, there is nothing left to prove that he communicated his thought in writing before 1620, making it impossible to track his ideas along any timeline. This could reflect a new intellectual maturity or a growing sense of personal and professional security, connected in particular, perhaps, to the 1619 appointment of the anti-Calvinist John Howson to the diocese of the bishop of Oxford. It might also have to do with the concrete angelic insights he received about theological matters throughout the late 1610s and early 1620s; see chapter 3 on his converse with angels.

6. Such search for truth required the diligent study of God's two books, nature and scripture, as mutually illuminating. On the metaphor of "the two books of God," see Brooke, *Sci-*

ence and Religion; Crocker, *Religion*; Harrison, *Bible and Fall of Man*; Killeen and Forshaw, *Word and the World*; Van der Meer and Mandelbrote, *Nature and Scripture*.

7. A full discussion of Napier's theological outlook is beyond the scope of this book. For an interesting discussion of his letter about the lawfulness of usury, together with its full transcription, see MacDonald, "Defence of Usury."

8. In light of recent historiography, I have exercised extra caution in using the terms "papist" (or "popish") and "Puritan," both loaded terms of abuse and yet still helpful in portraying seventeenth-century English polemics. Like many other men within the Church of England, Napier used the terms "papist" and "popish" indiscriminately to denote people, beliefs, and practices suspected as being tied or related to the Church of Rome. I follow this line in my use of the two terms. Following Patrick Collinson, Peter Lake, Anthony Milton, and others, I employ the term "Puritan" to denote Protestants "distinctive in their enthusiasm and zeal for the cause of true religion" (Milton, *Catholic and Reformed*, 8), both in their own eyes and in the eyes of others. I use "conservative" to refer to those clergymen who did not share such attitudes. In most places I prefer the use of "pro-Calvinist" and "anti-Calvinist" as coordinates against which Napier's positions can be measured, the first denoting one's general sympathy toward further reformation in purely doctrinal issues (in particular the doctrine of predestination) and the second its antonym.

9. "They are not of the same bodye that we are," he proclaimed of the Roman Catholics, "for they hold & have another head then we have." Ashm. 1473, fol. 454. This quote is taken from Napier's "anti-papist" exposition on "wheather the popish church be quite broken of[f] from being a [church *lost in binding*]." Ashm. 1473, fols. 452–56. Similar rhetoric is expressed in a very short text titled "De fide pontificioru[m]." Ashm. 1473, fol. 451.

10. Ashm. 1473, fol. 451.

11. See his letter on free will: Napier to [Sir Robert Napier?], date unknown, Great Linford, Ashm. 1730, fols. 212–13v. An incomplete draft of the same letter is in Ashm. 1730, fols. 203r–v.

12. Napier to William Page, October 22, 1629, Great Linford, Ashm. 1730, fols. 224r–v. Page, a doctor of divinity and fellow of All Souls College, was also Napier's patient. In one of his queries to the archangel Raphael, the doctor noted that Page was "crazed with study." Ashm. 223, fol. 192v.

13. See my discussion below on citing pagan testimonies in a church sermon.

14. Ashm. 1473, fol. 452.

15. Ashm. 1473, fol. 453.

16. For a somewhat similar contemporary outlook, see Archbishop Abbot's letter from 1622, in which he condemns both "popery" and popularity, denouncing "the superstition of the one and the madness of the other." Abbott, "Archbishop Abbot's Letter." Bernard Capp, who studied the life of the poet John Taylor, found that he accepted the doctrine of predestination, admired Calvin and Beza, and disapproved of Arminians but, at the same time, thought that people were free to accept or reject God's grace, that prayers could help relieve a sinner's soul, and that good works were inseparable from faith. People like Taylor, suggests Capp (in my view unconvincingly), "were never required to think through their ideas or to fashion them into a coherent whole, . . . [and] were likely to combine clearly Protestant beliefs with others seemingly incompatible." Capp, *John Taylor*, 133–35. See also Peter Elmer on the famous Anglican physician Thomas Browne: Elmer, "Medicine, Religion," 35–36.

17. See, for example, Jewel, *Apology*.

18. See, for example, William Page's unpublished treatise, arguing that "the catholic rule of expounding Scripture" should rely on the consensus of the church fathers of the first four centuries. The Queen's College, Oxford, MS 247, fols. 18–41v, cited in Quantin, *Church of England*, 192. Page corresponded with Napier on theological matters.

19. See my discussion of his "Defence of Astrology" in chapter 1.

20. Avis, *In Search of Authority*, 77–82.

21. "The Thirty-Nine Articles, 1563," cited in Cressy and Ferrel, *Religion and Society*, 75.

22. Napier to an unknown recipient, date unknown, Great Linford, Ashm. 1730, fol. 206.

23. This decree, known as *Firmiter Credimus*, implied that angels were not corporeal in any way. Tanner, *Decrees*, 1:230. For a late seventeenth-century discussion (first published from manuscript in 1879) of the discrepancy between the conclusions of the Second Nicene and Fourth Lateran Councils on the corporality of angels, see Sinistrari, *Demoniality*, 73–79.

24. The *Cantate Domino*, the Papal Bull of Pope Eugene IV issued at the Council of Florence, confirmed the Roman-Catholic Canon of the Bible, including the forty-six books of the Old Testament and twenty-seven of the New Testament, which Pope Damasus I had published a thousand years earlier at the Synod of Hippo (393) and which would later be repeated by the Council of Trent.

25. Ashm. 1730, fol. 206.

26. Napier to Henry Jackson, June 4, 1621, Great Linford, Ashm. 1730, fols. 180–81. Jackson was admitted to Corpus Christi College at the age of sixteen, received his MA, and was elected fellow in 1612. He was employed by John Spenser in the 1610s to prepare several of Richard Hooker's unpublished works for print (these included some of Hooker's tractates and sermons, but not his *Laws of Ecclesiastical Polity*). Jackson was also involved in translating some treatises of other English theologians into Latin, including those of John Reynolds, Sebastian Benefield, and William Whitaker. Eppley, "Jackson, Henry."

27. Napier to an unknown recipient, date unknown, Great Linford, Ashm. 1730, fols. 182–83.

28. Evans, *Problems of Authority*; Jenkins and Preston, *Biblical Scholarship*.

29. A critical view of the textual tradition of the scriptures was not a Protestant monopoly in the seventeenth century. See Malcolm, *Aspects of Hobbes*, 383–431.

30. See his two letters from 1525 to the Christians of Antwerp and against "the murdering prophets" who led the Peasants' War. Luther, "Against the Robbing."

31. Evans, *Problems of Authority*, 27–29.

32. The Erasmian term *adiaphora*, or "things indifferent," refers to matters not regarded as essential to faith. The issue of what constituted *adiaphora* was widely disputed during the Reformation, most notably between Erasmus and Luther. For more on the English debate over "probable knowledge" versus "infallible certainty" in religion, see Shapiro, *Probability*, 78–82; *Culture of Fact*, 168–88; Van Leeuwen, *Problem of Certainty*.

33. For instance, in a question posed to the archangel Raphael, Ashm. 237, fol. 192v.

34. On the authority of tradition in English polemics, see, for example, Quantin, *Church of England*, 191ff.

35. Ashm. 1730, fol. 184.

36. Ashm. 1730, fol. 206.

37. Ashm. 1730, fol. 182v; see also fol. 180v. A most curious short exposition in Napier's handwriting, bound in the same casebook, contrasts, from an allegedly neutral position, the Roman Catholic and Protestant claims to absolute authority (Ashm. 1730, fols. 218r–v). Advocating a rejection of the Protestant sole reliance on scripture and the Anglican Sixth Article ("Of the Sufficiency of the Holy Scriptures for Salvation"), it denies the "ability or meenes in the protestant church to have sutch divine & infallible knowledge & assuranc of religion," before suddenly ending in midsentence. The author declares that if proof is given that "ther[e] is any sutch divine or infallible meanes in the protestant churche," he shall willingly submit himself to its authority. A standard trope in post-Reformation controversy, such phrasing would have been used only by a convinced Roman Catholic. More than likely, therefore, the text is not Napier's original; it may have been transcribed by the doctor because the arguments within it chimed with some of his own doubts about certain aspects of Protestant doctrine, especially *sola scriptura*. I thank Prof. Anthony Milton for sharing with me his thoughts on this puzzling short text.

38. "The unanimous consent of the Fathers" had a further specific meaning at the Council of Trent, where it was applied particularly to the church fathers' interpretation of scripture. The council reconfirmed the Roman Catholic assertion that the key for the proper understanding of scripture was held solely by the authority of the church and the church fathers.

39. Ashm. 1148, p. 207. Cf. Ashm. 1730, fols. 182r–v.

40. Ashm. 1730, fol. 213v.

41. Ashm. 1148, p. 204.

42. Ashm. 1730, fol. 180v. The closing phrase, which translates as "upon analogy of faith," was used in exegesis to denote "scripture explaining scripture."

43. Ashm. 1730, fol. 180. "The catholick . . . church" should be read here as "the universal church."

44. Ashm. 1730, fol. 182v.

45. Ashm. 1730, fol. 180.

46. Ashm. 1730, fol. 180v.

47. Ibid.

48. Ashm. 1730, fol. 180.

49. Ashm. 1148, pp. 203–28, here 227.

50. Black, *Catalogue*, 1008.

51. Bull, *Newport Pagnell*, 107.

52. For James's handwriting, see his letter to Napier on Ashm. 240, fol. 147 and his copies of "Liber Salamonis" and "Liber Lunæ" in Sloane 3826.

53. Particularly similar is Napier's "Defence of Astrology," in which he cites, in the same meticulous and repetitive manner, and with a very similar wording, dozens of fragmentary references to support his position. Ashm. 204, fols. 50–63v; Ashm. 240, fols. 137–38v. See my discussion of this text in chapter 1.

54. Arguments to the contrary might be Lilly's aforementioned testimony that Napier did not preach (while the tract is based on a sermon he delivered) and the fact that the tract is found in Ashm. 1148 rather than in Ashm. 1730, the case in which most of Napier's theological writings were gathered. The tract's antagonistic tone, as I have just noted, is also uncharacteristic of Napier.

55. Ashm. 1148, p. 203.

56. The identity of Napier's adversary could not be verified. According to both Black and Bull it was one of the vicars of Newport Pagnell, but this does not accord with the apparent awe that Napier continuously displays toward his accuser. Another alternative is William Twisse: a distinguished pro-Calvinist clergyman who was the vicar of Newbury and later the prolocutor of the Westminster Assembly. In June 1617, Napier remarked in his diary, "Doctor Twyst [sic] & master Tyringham assaulted me in disputat[ion] & d[octor] T scoffed at me." Ashm. 220, fol. 40v. I did not find sufficient proof for either alternative.

57. The quote, recited in full (in Latin) in Napier's tract, is taken from Horace's *Ars poetica*. A translated and edited version can be found in Horace, *Collected Works*, 292–93.

58. Clear references to a written attack on Napier can be found on Ashm. 1148, pp. 203–4.

59. Ashm. 1148, p. 204.

60. Hughes, "Religious Polemic." Hughes has also shown how in such incidents the spoken, written, and printed word were often intertwined (as we can see in this particular case).

61. See note 8 above.

62. Lake, "Defining Puritanism," 15.

63. Lake, *Anglicans and Puritans*, 27; Milton, *Catholic and Reformed*, 537.

64. Ashm. 1148, p. 203. Here, too, Napier uses classical imagery, probably to further annoy his opponent.

65. "Neither giue heed to fables, and endlesse genealogies, which minister questions, rather then edifying which is in faith: so doe," 1 Timothy 1:4; "But refuse prophane and olde wiues fables, and exercise thy selfe rather vnto godlinesse," 1 Timothy 4:7; "And they shall turne away their eares from the trueth, and shall be turned vnto fables," 2 Timothy 4:4.

66. Babington, *Sermon*, 25–26. Napier refers directly to this story in Babington's court sermon, in Ashm. 1148, p. 219. I found the character of Nephastes untraceable, perhaps a distorted name of some Egyptian goddess.

67. Seznec, *Survival*, 84–99; Press, "Subject and Structure."

68. A pioneering work in the Christian context was the *Mythologiae* of the sixth-century writer Fulgentius Planciades, suggesting an allegorical rendering of many classical myths.

69. Jeanneret, "Renaissance Exegesis." Some church fathers, however, warned against such interpretations, which they feared could blur the distinction between Christian and pagan cultures.

70. Robert Burton, in his *Anatomy of Melancholy*, relies on examples from both scripture and pagan sources in demonstrating how God can sometimes cause disease. Burton, *Anatomy of Melancholy*, 2:159–60.

71. Killeen, *Biblical Scholarship*, 102–8; Williams, *Common Expositor*, 199–215. Henry Ainsworth, a biblical commentator who was Napier's contemporary, supported his allusions to heathen testimonies "for the witnesse which they beare unto the truth of God." Ainsworth, preface to *Annotations*, 7. The seventeenth-century English theologian Theophilus Gale suggested that many of Homer's fictions are based on real scriptural traditions that Homer had gathered while in Egypt. For a similar suggestion, see Fludd, *Mosaicall Philosophy*, 41–44. Napier himself, in his "Defence of Astrology," reiterates the widely accepted opinion that it was none other than Abraham who had taught astrology to the Egyptian priests at Heliopolis. Ashm. 242, fol. 187v.

72. The notion of a common and universal basis for all religions is known as *prisca theologia* or *theologia perennis*. See Stroumsa, *New Science*.

73. Hanegraff, *Esotericism*, 41–68. For Ficino's belief that ancient poetry was a genuine source of revelation, see Allen, *Synoptic Art*, 25.

74. Napier's tract is not a biblical commentary and is evidently based on a few comments he made in a church sermon. Since we have almost no record of that sermon, a direct discussion of it as such is not possible. It should be kept in mind, nonetheless, that the sermon is by definition a freer genre. For the role of the sermon in English seventeenth-century ecclesiastical and political controversies, see the collection of articles in Ferrell and McCullough, *English Sermon*.

75. Ashm. 1148, p. 222, based on Nehemiah 8:8.

76. Ashm. 1148, p. 209; see also p. 225.

77. Ashm. 1148, p. 208.

78. Ashm. 1148, p. 204. On pagan gods as representations or transformations of biblical heroes, see Stroumsa, *New Science*. Napier also invokes Apollos of Alexandria, who was "shewing by the scriptures that Christ was Jesus." Ashm. 1148, p. 222.

79. Ashm. 1148, p. 219.

80. Ashm. 1148, p. 208.

81. Ashm. 1148, p. 224.

82. Ashm. 1148, p. 205.

83. Ashm. 1148, p. 226.

84. Ashm. 1148, p. 221.

85. Ashm. 1148, p. 225. Napier repeatedly uses this phrase as a general title for the accusation made against him.

86. McGrath, *Intellectual Origins*, chap. 6; Harrison, *Bible*. On the rejection of allegorical interpretations by Calvin, see Thompson, "Calvin"; McKim, *Calvin*.

87. Stroumsa, *New Science*; Williams, *Common Expositor*, 255–68. The incorporation of some parts of pagan wisdom, however, especially philosophy, has sometimes cut across the religious divide. See, for example, Blair, "Mosaic Physics."

88. Harrison, "Hermeneutics."

89. Napier to an unknown recipient, January 12, 1629, Great Linford, Ashm. 1730, fols. 170r–v, here 170r. The Latin quote translates literally as "The sheep of St. Peter, to whom befell the duty to give pastoral care to these sheep of this shepherd."

90. Augustine, *Homilies*, 2:218 (Homily 14). Augustine's reasoning is based on his interpretation of "Hee must increase, but I must decrease," John 3:30. December 25, occurring four

days after the winter solstice, is almost exactly the time of year in which the length of days starts to increase. The four-day discrepancy was explained by the shift of the calendar from the actual astronomical pattern at Christ's birth to the fourth century.

91. Based on Luke 3:21 and onward.

92. In festivals such as *natalis solis invicti* (the Roman "birth of the unconquered Sun") and the birthday of Mithras.

93. Other arguments dealt with refuting one or more of the aforementioned calculations or questioned the possibility of Christ's birth taking place in the cold and rainy month of December.

94. Some sources from the second half of the seventeenth century, echoing the English debate on this issue, are *Against the Observation of a Day in Memory of Christs Birth* (London, 1660); *Christs Birth not Mis-Timed* (London, 1648); Collinges, *Responsoria*; John Taylor, *Christmas In & Out: Or, Our Lord & Saviour Christs Birth-Day* (London, 1653); Mather, *Testimony*; Robert Skinner, *Christs Birth Miss-Timed* (London, 1648).

95. Ashm. 1730, fol. 170.

96. Based on the events described in Luke 2:1–8.

97. In the contents of his celebrated work on probability, Ian Hacking has pointed out the earlier meaning of "probability," according to which a probable opinion was "one which was approved by some authority, or by the testimony of respected judges." Hacking, *Emergence*, xi.

98. Numbers 24:17.

99. Matthew 2:2.

100. Albertus Magnus, *Speculum Astronomiae*, 36–37; Renaker, "Horoscope."

101. Bacon, *Opus Majus*, 1:285–86.

102. North, *Horoscopes*, 163–73. For a recent discussion of the history of chronology through the dating of Christ's life events, see Nothaft, *Dating the Passion* and "From Sukkot."

103. The famous Italian astrologer Girolamo Cardano, taking for granted that he knew when Christ was born, disregarded the criticism and placed Christ's horoscope at the very heart of his work. He even used Christ's nativity to prove Ptolemaic astrology, pointing out that it accurately foretold the Savior's life events. Grafton, *Cardano's Cosmos*, 151–55.

104. Ashm. 1730, fol. 170. This accords with Christ's nativity chart drawn by Pierre d'Ailly, in which the Ascendant is 8° into Virgo and the Sun is in the fifth house, which is 1° into Capricorn. Smoller, *History*, 18, reproduced from Pierre d'Ailly, *Tractatus de Imagine Mundi et Varia Ejusdem Auctoris et Joannis Gersonis Opuscula* (Louvain, ca. 1483), fol. ee2v. About charting Christ's nativity, see also Geneva, *Astrology*, 155–56.

105. Collinges, *Responsoria*, 70.

106. Mather, *Testimony*, 21. The idea that the natural world should be interpreted in tandem with scripture was criticized by some Reformers, who insisted on the sufficiency of the Bible for the Christian faith. Harrison, "Hermeneutics," 353.

107. Ashm. 242, fol. 189v. Richard Carpenter, an English clergyman and theological author, wrote in his 1657 defense of astrology, "I prove it upon this account, that scripture cannot be rightly understood without enquiry made into Astrology. It is written in the prophet Amos, where Moloch is introduced: the star of your God, which ye made to your selves. It is necessary, for the right and full understanding of this place, that it be known what star this was, what star in heaven it resembled, and what likewise is the nature, power and work of that star." Carpenter, *Astrology*, 17.

108. In his study of Thomas Browne, Kevin Killeen identifies a similar process in which physical explanations and biblical exegesis acted as "mutually reinforcing discourses." Killeen, *Biblical Scholarship*, 127, 142–43.

109. Grafton, *Defenders of the Text*, 104–44. On the contemporary chronology of ancient events, see also Killeen, *Biblical Scholarship*, 90–101.

110. Smoller, "Astrology."

111. The Latin quote translates literally as "Everyone is to be believed (said Aristotle) in reference to his own art or profession." Ashm. 1730, fol. 170v.

112. Most of the new philosophies proposed in this period, including Platonism, Paracelsianism, and experimental natural philosophy, were claimed as more pious or Christian than the older Aristotelian philosophy. Blair, "Mosaic Physics," 33.

CONCLUSION

1. Ash, *Power*, 213–16.
2. Harrison, *Bible* and "Hermeneutics."
3. Ashm. 414, fol. 47.
4. To say that Napier was experimenting is to state a fact; it is not an ascription of modernity or of participation in early modern science. It is, for instance, rather clear that his boundaries of error and truth were not determined by methodological experimentation.
5. I contrast Napier's outlook with the "mechanistic worldview," although he too incorporated mechanistic metaphors such as the clockwork cosmos, as in his "Defence of Astrology." Ashm. 242, fol. 190.
6. It has already been acknowledged, however, that no clear-cut division existed between those who were still in awe of ancient authority and those who rejected all old knowledge. For a comprehensive discussion of the seventeenth-century shift from "ancient" to "new" knowledge, see Shapin, *Scientific Revolution*, 65–80.
7. For some exemplary case studies of early modern correspondence, see Botley and Van Miert, *Correspondence*; Hunter, Clericuzio, and Principe, *Correspondence*; Hall and Hall, *Correspondence*.
8. Shapin, *Scientific Revolution*, esp. 30–46.
9. It might be odd to think of Napier as a promoter of new science, unless we sincerely accept that the story of science included historical elements that were eventually abandoned and have now faded into oblivion.

BIBLIOGRAPHY

PRIMARY SOURCES IN MANUSCRIPT

Manuscripts pertaining to Richard Napier, Ashmolean Collection, Bodleian Library, Oxford

For further bibliographical information on the Ashmolean Collection, see William H. Black, *A Descriptive, Analytical, and Critical Catalogue of the Manuscripts Bequeathed unto the University of Oxford by Elias Ashmole* (Oxford: Oxford University Press, 1845); and W. D. Macray, *Index to the Catalogue of the Manuscripts of Elias Ashmole* (Oxford: Oxford University Press, 1866).

174	Astrological-medical practice 1599–1631 (loose notes); medical collections, notes, and recipes; astrological calculations and notes; correspondence to and from Napier
175	Astrological-medical practice 1597; theological papers and notes; astrological calculations and notes
177	Alchemical papers and notes; astrological calculations and notes; correspondence to Napier
181	Astrological-medical practice 1598–1629 (loose notes); observations and rules for the practice of astrological medicine; drafts of sermons; medical collections, notes, and recipes; correspondence to and from Napier
182	Astrological-medical practice 1597–1598; astrological calculations and notes; geomantic and onomantic calculations; prayers and exorcisms; drafts of sermons; theological papers and notes
188	Astrological calculations and notes
193	Astrological-medical practice 10/1606–1/1608
194	Astrological-medical practice 12/1629–6/1630; medical collections, notes, and recipes
196	Astrological-medical practice 4/1615–2/1616; questions proposed to angels
197	Astrological-medical practice 2/1603–2/1604
198	Astrological-medical practice 9/1616–6/1617; medical collections, notes, and recipes; questions proposed to angels
199	Astrological-medical practice 12/1612–8/1613
200	Astrological-medical practice 1/1611–2/1612; astrological calculations and notes; questions proposed to angels
201	Astrological-medical practice 2/1618–8/1618; questions proposed to angels; medical collections, notes, and recipes
202	Astrological-medical practice 2/1600–12/1600; medical collections, notes, and recipes
203	Astrological-medical practice 4/1609–3/1610
204	Astrological-medical practice 1599–1619 (loose notes); theological papers and notes; medical collections, notes, and recipes; astrological calculations and notes; a treatise defending astrology; correspondence to and from Napier

BIBLIOGRAPHY

207	Astrological-medical practice 2/1603–4/1605
211	Astrological-medical practice 4/1633–10/1633; medical collections, notes, and recipes
212	Astrological-medical practice 7/1631–3/1632; medical collections, notes, and recipes
213	Astrological-medical practice 9/1619–4/1620; autobiographical notes; medical collections, notes, and recipes; questions proposed to angels
214	Astrological-medical practice 9/1632–4/1633; medical collections, notes, and recipes; astrological calculations and notes
215	Astrological-medical practice 3/1605–4/1606 (some later)
216	Astrological-medical practice 3/1605–12/1605
217	Astrological-medical practice 4/1625–8/1625; medical collections, notes, and recipes
218	Astrological-medical practice 3/1623–10/1623; prayers and exorcisms; medical collections, notes, and recipes; questions proposed to angels
220	Astrological-medical practice 6/1617–2/1618; theological notes; questions proposed to angels
221	Astrological-medical practice 3/1602–2/1603; medical collections, notes, and recipes
222	Astrological-medical practice 7/1622–2/1623; medical collections, notes, and recipes; astrological calculations and notes; notes concerning Napier's ordination; theological papers and notes; questions proposed to angels
223	Astrological-medical practice 2/1622–7/1622; medical collections, notes, and recipes; questions proposed to angels
224	Astrological-medical practice 9/1625–2/1627; medical collections, notes, and recipes; astrological calculations and notes
227	Astrological-medical practice 2/1627–11/1627
228	Astrological-medical practice 7/1598–2/1600; medical collections, notes, and recipes; astrological calculations and notes
229	Astrological-medical practice 1/1608–4/1609
230	Astrological-medical practice 8/1618–5/1619; questions proposed to angels
231	Astrological-medical practice 6/1621–2/1622; questions proposed to angels
232	Astrological-medical practice 11/1630–7/1631
233	Astrological-medical practice 11/1620–5/1621; medical collections, notes, and recipes
234	Astrological calculations and notes
235	Astrological-medical practice 5/1619–9/1619; questions proposed to angels
237	Astrological-medical practice 5/1614–4/1615; medical collections, notes, and recipes; theological papers and notes; invocations of angels; questions proposed to angels
238	Astrological-medical practice 6/1630–11/1630; astrological calculations and notes; theological papers and notes
239	Astrological-medical practice 3/1610–1/1611
240	Astrological-medical practice 1599–1624 (loose notes); medical collections and notes; astrological calculations and notes; a treatise defending astrology; correspondence to and from Napier
242	A treatise defending astrology
243	Astrological calculations and notes; medical recipes; correspondence to and from Napier
244	Invocations of angels
313	A nativity of King James I
329	Astrological-medical practice 11/1609–11/1610; theological papers and notes
334	Astrological-medical practice 3/1610–5/1611

335	Astrological-medical practice 2/1609–9/1610 (some later)
338	Astrological-medical practice 9/1608–2/1609; theological papers and notes; drafts of sermons
339	Astrological calculations and notes
356	Astrological calculations and notes
402	Astrological-medical practice 6/1624–3/1625; questions proposed to angels
403	A book of medicine and astrology, transcribed from Simon Forman's MSS
404	Astrological-medical practice 1/1601–3/1602; medical collections, notes, and recipes
405	Astrological-medical practice 4/1623–12/1628
406	Astrological-medical practice 5/1629–12/1629; medical collections, notes, and recipes
407	Astrological-medical practice 12/1628–5/1629; medical collections, notes, and recipes
408	Astrological-medical practice 2/1616–9/1616; questions proposed to angels
409	Astrological-medical practice 2/1612–12/1612; invocations of angels; questions proposed to angels
410	Astrological-medical practice 11/1627–1/1628
412	Astrological-medical practice 10/1633–4/1634
413	Astrological-medical practice 10/1623–6/1624; autobiographical notes; questions proposed to angels
414	Astrological-medical practice 4/1620–11/1620; astrological calculations and notes; questions proposed to angels
415	Astrological-medical practice 3/1604–3/1605; alchemical papers and notes; observations and rules for the practice of astrological medicine
416	Astrological-medical practice 3/1632–9/1632
417	Astrological calculations and notes
421	Prayers and exorcisms; correspondence to and from Napier
423	Theological papers and notes
431	Correspondence to and from Napier
546	Astrological calculations and notes
752	Books and manuscripts owned and annotated by Napier
1148	An apology for using the learning of the gentiles
1283	Prayers and exorcisms
1293	License to practice medicine
1298	Notes concerning Napier's induction to Great Linford
1386	Medical collections, notes, and recipes
1388	Medical collections, notes, and recipes
1399	Medical collections, notes, and recipes
1400	Alchemical papers and notes
1407	Alchemical papers and notes
1413	Medical collections, notes, and recipes
1414	Correspondence to and from Napier; medical collections, notes, and recipes; astrological calculations and notes; notes and miscellanies
1421	Alchemical papers and notes; books and manuscripts owned and annotated by Napier
1423	Alchemical papers and notes
1424	Alchemical papers and notes
1429	Alchemical papers and notes; books and manuscripts owned and annotated by Napier
1437	Medical collections, notes, and recipes
1441	Alchemical papers and notes
1442	Medical collections, notes, and recipes

1447	Medical collections, notes, and recipes
1453	Medical collections, notes, and recipes; magical charms
1457	Alchemical papers and notes; medical collections, notes, and recipes
1458	Alchemical papers and notes; correspondence to and from Napier
1473	Medical notes and recipes; theological papers and notes; drafts of sermons; alchemical papers and notes
1478	Medical collections, notes, and recipes
1480	Medical collections, notes, and recipes
1483	Books and manuscripts owned and annotated by Napier
1484	Books and manuscripts owned and annotated by Napier
1485	Books and manuscripts owned and annotated by Napier
1488	Medical collections, notes, and recipes; astrological calculations and notes; magical notes; correspondence to and from Napier
1490	Alchemical papers and notes
1492	Alchemical papers and notes
1497	Medical collections, notes, and recipes
1501	Astrological calculations and notes; correspondence to and from Napier
1730	Alchemical papers and notes; theological papers and notes; prayers; correspondence to and from Napier
1790	Prayers, charms, and exorcisms; invocations of angels

Manuscripts pertaining to Richard Napier, Simon Forman, and Elias Ashmole, Sloane Collection, British Library, London

For further bibliographical information on the Sloane Collection, see Edward J. L. Scott, *Index to the Sloane Manuscripts in the British Museum* (London: British Museum, 1904).

519	Medical recipes
1087	Medical recipes
2550	Forman's observations and rules for the practice of astrological medicine
3679	Books and manuscripts owned and annotated by Napier; Napier's extracts from *The Picatrix*
3822	Notes about magic and astrology; prayers and exorcisms; Napier's apology for the use of images; a short prophecy by the archangel Raphael; correspondence to and from Napier
3826	Books and manuscripts owned by Napier
3846	Notes about magic and astrology; invocations of angels
3854	Notes about magic, astrology, and religion; papers that Napier wrote or owned
3884	Papers that Napier wrote or owned

Manuscripts pertaining to Richard Napier and Elias Ashmole, Additional Manuscripts, British Library, London

36674	Notes about magic and astrology; prayers and invocations of angels; Ashmole's account of the case of Elizabeth Jennings

Manuscripts pertaining to Richard Napier, Corpus Christi College, Oxford

168	Observations and rules for the practice of astrological medicine
169	Observations and rules for the practice of astrological medicine
170	Observations and rules for the practice of astrological medicine

Manuscripts pertaining to Richard Napier, Stowe Collection, Huntington Library, San Marino, Calif.

1500	Napier's letter about the voidance of blood
1501	Napier's letter of recommendation
1502	Napier's letter of recommendation

Manuscripts of Simon Forman, Ashmolean Collection, Bodleian Library, Oxford

219	Astrological-medical practice 2/1599–12/1599
234	Astrological-medical practice 3/1596–2/1597
236	Astrological-medical practice 1/1600–12/1600
240	Astrological calculations and notes; correspondence to and from Forman
244	Forman's book of Kabbala and angels; astrological collections and notes; medical notes
354	A book of geomancy; onomantic notes
355	A book of medicine and astrology
389	A book of medicine and astrology
395	A book of medicine and astrology
411	Astrological-medical practice 12/1600–11/1601 and 05/1603–09/1603
431	Medical and magical notes
1488	Medical recipes; medical and magical notes; correspondence to and from Forman
1495	Observations and rules for the practice of astrological medicine

Manuscripts of Sir Richard Napier, Ashmolean Collection, Bodleian Library, Oxford

363	A book of medicine and astrology, transcribed from Simon Forman's MSS
412	Astrological-medical practice 04/1634–12/1634
1457	Medical recipes annotated by Sir Richard Napier
1488	Medical collections and notes

Manuscripts pertaining to Elias Ashmole, Ashmolean Collection, Bodleian Library, Oxford

374	Documents relating to Ashmole's life and practice
421	Notes on sigils, extracted from Napier's books; William Lilly's autobiography
423	List of nativities in Napier's books; selective list of diseases and matters in Napier's books
1790	Notes of diseases and medicines, extracted from Napier's books; correspondence to and from Ashmole

Other Manuscripts, Ashmolean Collection, Bodleian Library, Oxford

183	John Booker's astrological practice
339	John Booker's astrological practice

Other Collections, Bodleian Library, Oxford

Aubrey 5	John Aubrey's collections and notes
Aubrey 6	John Aubrey's collections and notes
Tanner 456a	John Aubrey's letter to Anthony Wood

Documents, Centre for Buckinghamshire Studies, Aylesbury

D-U   Deeds, accounts, and other records relating mainly to the Great Linford and Lathbury estates of the Uthwatt and Andrewes families

Documents, Milton Keynes Local Studies and Family History Library, Milton Keynes
Hanslope Church Register for 1571-1732 (microfilm)

PRINTED PRIMARY SOURCES

Abbott, George. "Archbishop Abbot's Letter regarding Preaching." In *Records of the Old Archdeaconry of St. Albans: A Calendar of Papers AD 1575 to AD 1637*, edited by H. R. Wilton Hall, 150–52. St. Albans: St. Albans and Hertfordshire Architectural and Archaeological Society, 1908.

Agrippa, Henry Cornelius. *Three Books of Occult Philosophy*. Edited by Donald Tyson. Translated by James Freake. St. Paul, Minn.: Llewellyn, 1993.

Ainsworth, Henry. *Annotations upon the First Book of Moses, Called Genesis*. London, 1627.

Albertus Magnus. *Speculum Astronomiae*. Edited by Stefano Caroti, Michela Pereira, Stefano Zamponi, and Paola Zambelli. Pisa: Domus Galilaeana, 1977.

Anderson, Roger C., ed. *The Book of Examinations and Depositions, 1622-44*. Southampton: Southampton Rec. Soc., 1931.

Aquinas, Thomas. *De Occultis Operibus Naturae*, with a commentary. Edited and translated by Joseph B. McAllister. Washington, D.C.: Catholic University of America Press, 1939.

———. *Summa Theologiae*. Vol. 9, *Angels*, edited and translated by Kenelm Foster. Cambridge: Blackfriars, Eyre and Spottiswoode; New York: McGraw-Hill, 1968.

Ashmole, Elias. *Theatrum Chemicum Britanicum*. London, 1652.

Atwell, George. *An Apology: Or, Defense of the Divine Art of Natural Astrology*. London, 1660.

Aubrey, John. *Aubrey's Brief Lives*. Edited by Oliver L. Dick. 1949. Reprint, London: Secker and Warburg, 1960.

———. *Miscellanies upon Various Subjects*. London, 1696.

Augustine. "De doctrina christiana." In *Aurelii Augustini opera*, edited by J. Martin, 1–167. Corpus Christianorum, Series Latina 32. Turnhout: Typographi Brepols Editors Pontificii, 1962.

———. *Homilies on the Gospel According to St. John*. Translated by Henry Browne. Vol. 1 of 2. Oxford: John Henry Parker and F. and J. Rivington, 1848.

Aylmer, John. *An Harborovve for Faithfvll and Trevve Svbiectes*. London, 1559.

Babington, Gervase. *A Sermon Preached at the Covrt at Greenwich, the xxiiii of May, 1591*. London, 1591.

Bacon, Francis. *The Works of Francis Bacon, Baron of Verulam*. Vol. 1 of 5. London, 1765.

Bacon, Roger. *The "Opus Majus" of Roger Bacon*. Translated by Robert B. Burke. Vol. 1 of 2. 1928. Reprint, Philadelphia: University of Pennsylvania Press, 2000.

Balme, Maurice, ed. *Two Antiquaries: A Selection from the Correspondence of John Aubrey and Anthony Wood*. Edinburgh: Durham Academic Press, 2001.

Barrough, Philip. *The Method of Phisick*. London, 1590.

Biringuccio, Vannoccio. *De la Pirotechnia Libri X*. Edited and translated by Cyril S. Smith and Martha T. Gnudi. Cambridge, Mass.: MIT Press, 1966. Originally published in 1540 in Italian.

Black, William H. *A Descriptive, Analytical, and Critical Catalogue of the Manuscripts Bequeathed unto the University of Oxford by Elias Ashmole*. Oxford: Oxford University Press, 1845.

Blagrave, Joseph. *Blagrave's Astrological Practice of Physick*. London, 1672.

———. *Blagrave's Introduction to Astrology*. London, 1682.

Bliss, Philip. "Additions to Lord Orford's Royal and Noble Authors." *London Magazine* 5, no. 2 (1822): 387–90.

Botley, Paul, and Dirk van Miert, eds. *The Correspondence of Joseph Scaliger*. 8 vols. Geneva: Droz, 2012.

Britton, John. *Memoire of John Aubrey, F. R. S., Embracing his Autobiographical Sketches, a Brief Review of his Personal and Literary Merits, and an Account of his Works*. London: J. B. Nichols and Son, 1845.

Bull, Frederick W. *A History of Newport Pagnell*. Kettering: W. E. & J. Goss, 1900.

Burke, John. *A Genealogical and Heraldic History of the Commoners of Great Britain and Ireland*. Vol. 3 of 4. London: Henry Colburn, 1836.

Burke, John, and John B. Burke. *A Genealogical and Heraldic History of the Extinct and Dormant Baronetcies of England, Ireland and Scotland*. London: Scott, Webster and Geary, 1838.

Burton, Robert. *The Anatomy of Melancholy*. 2 vols. London, 1621.

Carpenter, Richard. *Astrology Proved Harmless*. London, 1657.

Casaubon, Meric. *A True and Faithful Relation of what Passed for Many Years between Dr. John Dee and Some Spirits*. London, 1659.

Casebooks Project (Welcome). Accessed March 25, 2017. http://www.magicandmedicine.hps.cam.ac.uk/.

"Classification of Diseases." *World Health Organization*. Modified November 26, 2016. http://www.who.int/classifications/icd/en/.

Cokayne, George E. *Some Notice of Various Families of the Name of Marsh*. Exeter: W. Pollard, 1900.

Coley, Henry. *Clavis Astrologiae Elimata: Or, A Key to the Whole Art of Astrology*. London, 1676.

Collinges, John. *Responsoria ad Erratica Piscatoris: Or, A Caveat for Old and New Prophanenesse*. London, 1653.

Collins, Arthur. *The English Baronetage: Containing a Genealogical and Historical Account of all the English Baronets Now Existing*. Vol. 1 of 2. London, 1741.

Cotta, John. *A True Discovery of the Empiricke with the Fugitive Physition and Quacksalver*. London, 1617. Originally published as *A Short Discoverie of the Unobserved Dangers of Severall Sorts of Ignorant and Unconsiderate Practisers of Physicke in England* (London, 1612).

Cressy, David, and Lori A. Ferrell. *Religion and Society in Early Modern England: A Sourcebook*. 2nd ed. New York: Routledge, 2005.

Culpeper, Nicholas. "A Catalogue of Simples in the New Dispensatory." In *Culpeper's Complete Herbal*, 255–375. London: Richard Evans, 1816.

———. *Culpeper's Astrologicall Judgment of Diseases*. London, 1655.

———. "Directions for Making Syrups, Conserves, etc. etc." In *Culpeper's Complete Herbal*, 199–211. London: Richard Evans, 1816.

———. *The English Physitian Enlarged*. 1652. Reprint, London, 1681.

———. "A Key to Galen's Method of Physic." In *Culpeper's Complete Herbal*, 376–98. London: Richard Evans, 1816.

———. *Semeiotica Uranica*. London, 1651.

Dee, John. *The Diaries of John Dee*. Edited by Edward Fenton. Charlbury: Day Books, 1998.

Del Río, Martín A. *Disquisitionum Magicarum*. Mainz, 1617.

Dunte, Ludwig. *Decisiones Mille et Sex Casuum Conscientiae*. Lubeck, 1664.

Dunton, John. *Dunton's Ghost*. London, 1714.

Eland, William. *A Tutor to Astrologie: Or, Astrologie Made Easy*. London, 1657.

Evans, John. *The Universall Medicine: Or, Vertues of the Antimoniall Cup*. London, 1634.

Ficino, Marsilio. *Three Books on Life: A Critical Edition and Translation with Introduction and Notes*. Edited and translated by Carol V. Kaske and John R. Clark. Binghamton, N.Y.: Medieval and Renaissance Texts and Studies, 1989.

Field, John. *Ephemeris Anni. 1557*. London, 1556.

Fludd, Robert. *Mosaicall Philosophy*. London, 1649.

Gadbury, John. *Natura Prodigiorum: Or, a Discourse Touching the Nature of Prodigies*. London, 1660.

———. *Thesaurus Astrologiæ: Or, An Astrological Treasury*. London, 1674.

Galen. "On the Causes of Disease." In *Galen on Food and Diet*, edited and translated by Mark Grant, 46–61. London: Routledge, 2000.

———. "On the Humours." In *Galen on Food and Diet*, edited and translated by Mark Grant, 14–18. London: Routledge, 2000.

Ginzberg, Louis. *The Legends of the Jews*. Translated from the German manuscript by Henrietta Szold and Paul Radin. 7 vols. Philadelphia: Jewish Publication Society of America, 1909–38. Reprint, Baltimore: Johns Hopkins University Press, 1998.

Godwin, William. *Lives of the Necromancers: Or, An Account of the Most Eminent Persons in Successive Ages, who have Claimed for Themselves, or to whom has been Imputed by Others, the Exercise of Magical Power*. London: Frederick J. Mason, 1834.

Grafton, Richard. *A Briefe Treatise Contayning Many Proper Tables and Easie Rules*. London, 1591.

Gunther, Robert T., ed. *Early Science in Oxford*. Vol. 14, *Life and Letters of Edward Lhwyd*. Oxford: Oxford University Press, 1945.

Hall, Alfred R., and Marie B. Hall, eds. and trans. *The Correspondence of Henry Oldenburg*. 13 vols. Madison: University of Wisconsin Press, 1965–86.

Hall, John. *Select Observations on English Bodies of Eminent Persons in Desperate Diseases*. Translated by James Cook. London, 1683.

Hall, Joseph. *Cases of Conscience Practically Resolved*. London, 1654.

Harms, Daniel, James R. Clark, and Joseph H. Peterson, eds. *The Book of Oberon: A Sourcebook of Elizabethan Magic*. Woodbury, Minn.: Llewellyn, 2015.

Hart, James. *The Arraignment of Urines*. London, 1623.

———. *KLINIKE: Or, the Diet of the Diseased*. London, 1633.

Hiebner, Israel. *Mysterium Sigillorum*. Translated by B. Clayton. London, 1698.

Horace. *Collected Works*. Translated by Lord Dunsany and Michael Oakley. London: Dent, 1961.

Horst, George C. *Zauber Bibliothek*. 6 vols. Mainz: Kupferberg, 1821–26.

Howell, James. *Epistolae Ho-Elianae: Familiar Letters Domestic and Forren*. London, 1645.

Jewel, John. *An Apology for the Church of England*. Edited by John E. Booty. Ithaca: Cornell University Press, 1963.

Joseph, Harriet. *Shakespeare's Son-in-Law: John Hall, Man and Physician*. Hamden: Archon Books, 1964.

Josten, C. H., ed. *Elias Ashmole (1617–1692): His Autobiographical and Historical Notes, His Correspondence, and Other Contemporary Sources Relating to His Life and Work*. 5 vols. Oxford: Clarendon Press, 1966.

Lambeth and the Vatican. Vol. 2 of 3. London: John Knight & Henry Lacey, 1825.

Lane, Joan. *John Hall and His Patients: The Medical Practice of Shakespeare's Son-in-Law*. Stratford-upon-Avon: Shakespeare Birthplace Trust, 1996.

L'Estrange Ewen, Cecil. *Witchcraft and Demonianism: A Concise Account Derived from Sworn Depositions and Confessions Obtained in the Courts of England and Wales*. London: Health Cranton, 1933.

Lilly, William. *Christian Astrology*. London, 1647.

———. *The Last of the Astrologers: Mr. William Lilly's History of His Life and Times from the Year 1602 to 1681*. Reprinted from the second edition of 1715 with notes and introduction by Katharine M. Briggs. London: Folklore Society, 1974.

Linton, Eliza L., ed. *Witch Stories*. 1861. Reprint, London: Chatto and Windus, 1883.
Lipscomb, George. *The History and Antiquities of the County of Buckingham*. Vol. 4 of 4. London: J. & W. Robins, 1847.
Luther, Martin. "Against the Robbing and Murdering Hordes of Peasants." In *Martin Luther*, edited by Ernest G. Rupp and Benjamin Drewery, 121–26. Documents of Modern History. London: Edward Arnold, 1970.
Lysons, Daniel. *The Environs of London*. Vol. 1 of 4. London, 1792.
Lysons, Daniel, and Samuel Lysons. *Magna Britannia*. 6 vols. London: T. Cadell and W. Davies, 1806–22.
Macray, W. D. *Index to the Catalogue of the Manuscripts of Elias Ashmole*. Oxford: Oxford University Press, 1866.
Madden, Samuel. *Memoirs of the Twentieth Century*. Vol. 1 of 6. London, 1733.
"Magna Britannia: Being a Concise Topographical Account of the Several Counties of Great Britain." *European Magazine and London Review* 54 (1808): 290–6.
Mather, Increase. *A Testimony Against Several Prophane and Superstitious Customs*. London, 1687.
[Moir, George]. "Demonology and Witchcraft." *Foreign Quarterly Review* 6 (1830): 1–47.
———. *Magic and Witchcraft*. London: Chapman and Hall, 1852.
Montague, Richard. *A New Gagg for an Old Goose*. London, 1624.
Napier, Mark. *Memoirs of John Napier*. Edinburgh: William Blackwood, 1834.
Noble, Mark. *The Lives of the English Regicides*. Vol. 2 of 2. London, 1798.
Pechey, John. *A Plain Introduction to the Art of Physick*. London, 1697.
Perkins, William. *A Salve for a Sicke Man*. 1595. Reprint, Cambridge, 1600.
Prideaux, John. *The Patronage of Angels: A Sermon Preached at the Court*. Oxford, 1636.
Primrose, James. *Popular Errours: Or, the Errours of the People in Physick*. Translated by R. Wittie. London, 1651.
Reynes, Robert. *The Commonplace Book of Robert Reynes of Acle. An Edition of Tanner MS 40*. Edited by Cameron Louis. New York: Garland, 1980.
"The Rosary." *Eccentricities of Literature and Life: Or, the Recreative Magazine* 1 (1821): 34–39.
Rutton, William L. *Three Branches of the Family of Wentworth*. Vol. 1, *Wentworth of Nettlestead*. London: Mitchell and Hughes, 1891.
Salkeld, John. *A Treatise of Angels*. London, 1613.
Salmon, William. *Synopsis Medicinæ: Or, A Compendium of Astrological, Galenical, & Chymical Physick*. London, 1671.
Saunders, Richard. *The Astrological Judgement and Practice of Physick*. London, 1677.
Scott, Walter. *Letters of Daemonology and Witchcraft*. London: John Murray, 1830.
———. *Secret History of the Court of James the First*. Vol. 2 of 2. Edinburgh: J. Ballantyne, 1811.
Seitz, Alexander. *Ein Nutzlich Regiment Wider Bie Bösen Frantzosen*. Pforzheim, 1509. Reprinted in *Alexander Seitz Sämtliche Schriften*, edited by Peter Ukena, vol. 1 of 5 (Berlin: Walter de Gruyter, 1970).
Sibbes, Richard. *Light from Heaven*. London, 1638.
Sinistrari, Ludovico M. *Demoniality: Or, Incubi and Succubi*. Paris: Isidore Liseux, 1879.
Smith, Robert C. *The Familiar Astrologer, and Easy Guide to Fate, Destiny, and Foreknowledge, by Raphael*. London: William Bennett, 1831.
Tanner, Norman, ed. *Decrees of the Ecumenical Councils*. Vol. 1 of 2. London: Sheed & Ward; Washington, D.C.: Georgetown University Press, 1990.
Tomlins, Thomas E., and John Raithby, eds. *The Statutes at Large, of England and Great-Britain: From Magna Carta to the Union of the Kingdoms of Great Britain and Ireland*. Vol. 4 of 20. London: G. Eyre and A. Strahan, 1811.
Turner, William. *A Compleat History of the Most Remarkable Providences*. London, 1697.
Wecker, Johann J. *Eighteen Books of the Secrets of Art & Nature*. Edited by R. Read. London, 1661.

Williams, Robert J. *St. Andrew's Church Great Linford: A Brief History*. Stantonbury: St. Andrew's Congregational Council, 1983.
Winthrop, John. *Winthrop Papers*. Vol. 1, *1498–1628*, edited by Worthington C. Ford. Boston: Massachusetts Historical Society, 1929.
Wirtzung, Christopher. *The General Practise of Physicke*. Translated by Iacob Mosan. London, 1617.
Wood, Anthony à. *Athenae Oxonienses: An Exact History of All the Writers and Bishops who Have Had Their Education in the University of Oxford*. 2 vols. London, 1692–93.
Wright, Thomas. *Narratives of Sorcery and Magic: From the Most Authentic Sources*. Vol. 2 of 2. London: Richard Bentley, 1851.
The Zohar. Translated by Harry Sperling and Maurice Simon. Vol. 1 of 5. 1931. Reprint, London: Soncino Press, 1973.

SECONDARY SOURCES

Akasoy, Anna, Charles Burnett, and Ronit Yoeli-Tlalim, eds. *Astro-Medicine: Astrology and Medicine, East and West*. Florence: Sismel, 2008.
Allen, Michael J. B. *Synoptic Art: Marsilio Ficino on the History of Platonic Interpretation*. Florence: Olschki, 1998.
Allen, Michael J. B., and Valery Rees, with Marin Davies, eds. *Marsilio Ficino: His Theology, His Philosophy, His Legacy*. Leiden: Brill, 2002.
Andrews, Jonathan. "Napier, Richard (1559–1634), Astrological Physician and Church of England Clergyman." In *Oxford Dictionary of National Biography*, edited by H. C. G. Matthew and B. Harrison, 40:181–83. New York: Oxford University Press, 2004.
Armstrong, David. "The Patient's View." *Social Science and Medicine* 18, no. 9 (1984): 737–44.
Ash, Eric H. *Power, Knowledge, and Expertise in Elizabethan England*. Baltimore: Johns Hopkins University Press, 2004.
Asprem, Egil. *Arguing with Angels: Enochian Magic and Modern Occulture*. Albany: State University of New York Press, 2012.
Aston, Margaret. *England's Iconoclasts*. Vol. 1 of 2. Oxford: Clarendon Press, 1988.
Avis, Paul. *In Search of Authority: Anglican Theological Method from the Reformation to the Enlightenment*. New York: Bloomsbury, 2014.
Azzolini, Monica. *The Duke and the Stars: Astrology and Politics in Renaissance Milan*. Cambridge, Mass.: Harvard University Press, 2013.
———. "Reading Health in the Stars: Prognosis and Astrology in Renaissance Italy." In *Horoscopes and Public Spheres: Essays on the History of Astrology*, edited by Günther Oestmann, H. Darrel Rutkin, and Kocku von Stuckrad, 183–205. Berlin: Walter de Gruyter, 2005.
Bailey, Michael D. *Magic and Superstition in Europe: A Concise History from Antiquity to the Present*. Lanham, Md.: Rowman and Littlefield, 2007.
Baker, Jim. *The Cunning Man's Handbook: The Practice of English Folk Magic, 1550–1900*. London: Avalonia, 2014.
Baldwin, Martha R. "Toads and Plague: Amulet Therapy in Seventeenth-Century Medicine." *Bulletin of the History of Medicine* 67, no. 2 (1993): 227–47.
Barry, Jonathan. *Raising Spirits: How a Conjuror's Tale Was Transmitted Across the Enlightenment*. Basingstoke: Palgrave Macmillan, 2013.
Beier, Lucinda M. *Sufferers and Healers: The Experience of Illness in Seventeenth-Century England*. London: Routledge, 1987.
Beitchman, Philip. *Alchemy of the Word: Cabala of the Renaissance*. Albany: State University of New York Press, 1998.
Bennett, Kate. "John Aubrey, Hint-Keeper: Life-Writing and the Encouragement of Natural Philosophy in the Pre-Newtonian Seventeenth Century." *Seventeenth Century* 22 (2007): 358–80.

Blair, Ann. "Mosaic Physics and the Search for a Pious Natural Philosophy in the Late Renaissance." *Isis* 91, no. 1 (2000): 32–58.
Bonzol, Judith. "The Medical Diagnosis of Demonic Possession in an Early Modern English Community." *Parergon* 26, no. 1 (2009): 115–40.
Boudet, Jean-Patrice. *Entre science et nigromance: Astrologie, divination et magie dans l'Occident médiéval (XIIe–XVe siècle)*. Paris: Publications de la Sorbonne, 2006.
Brach, Jean-Pierre. "Magic IV: Renaissance—17th Century." In *Dictionary of Gnosis and Western Esotericism*, edited by Wouter J. Hanegraaff, Antoine Faivre, Jean-Pierre Brach, and Roelof van den Broek, 731–38. Leiden: Brill, 2005.
Brooke, John H. *Science and Religion: Some Historical Perspectives*. Cambridge: Cambridge University Press, 1991.
Burnett, Charles. *Magic and Divination in the Middle Ages: Texts and Techniques in the Islamic and Christian Worlds*. Aldershot: Variorum, 1996.
Bylebyl, Jerome. "The Manifest and the Hidden in the Renaissance Clinic." In *Medicine and the Five Senses*, edited by W. F. Bynum and Roy Porter, 40–60. Cambridge: Cambridge University Press, 1993.
Cadose, Linda A. *The Hidden Chamber in the Great Sphinx*. Bloomington: AuthorHouse, 2012.
Calder, Ian R. F. "A Note on Magic Squares in the Philosophy of Agrippa of Nettesheim." *Journal of the Courtauld and Warburg Institutes* 12 (1949): 196–99.
Cameron, Euan. *Enchanted Europe: Superstition, Reason, and Religion, 1250–1750*. Oxford: Oxford University Press, 2010.
Campbell, Jane. *The Retrospective Review (1820–1828) and the Revival of Seventeenth-Century Poetry*. Toronto: Waterloo Lutheran University, 1972.
Capp, Bernard. *English Almanacs, 1500–1800: Astrology and the Popular Press*. Ithaca: Cornell University Press, 1979.
———. *The World of John Taylor the Water-Poet, 1578–1653*. Oxford: Clarendon Press, 1994.
Chapman, Allan. "Astrological Medicine." In *Health, Medicine, and Mortality in the Sixteenth Century*, edited by Charles Webster, 275–300. London: Cambridge University Press, 1979.
Christianson, Paul. *Reformers and Babylon: English Apocalyptic Visions from the Reformation to the Eve of the Civil War*. Toronto: University of Toronto Press, 1978.
Clark, Stuart. *Thinking with Demons: The Idea of Witchcraft in Early Modern Europe*. Oxford: Oxford University Press, 1997.
Clucas, Stephen. "False Illuding Spirits & Cownterfeiting Deuills: John Dee's Angelic Conversations and Religious Anxiety." In *Conversations with Angels: Essays Towards a History of Spiritual Communication, 1100–1700*, edited by Joad Raymond, 150–74. Basingstoke: Palgrave Macmillan, 2011.
Clulee, Nicholas H. *John Dee's Natural Philosophy: Between Science and Religion*. London: Routledge, 1988.
Condrau, Flurin. "The Patient's View Meets the Clinical Gaze." *Social History of Medicine* 20, no. 3 (2007): 525–40.
Cook, Harold J. *The Decline of the Old Medical Regime in Stuart London*. Ithaca: Cornell University Press, 1986.
———. *Matters of Exchange: Commerce, Medicine, and Science in the Dutch Golden Age*. New Haven: Yale University Press, 2007.
———. "The New Philosophy and Medicine in Seventeenth-Century England." In *Reappraisals of the Scientific Revolution*, edited by David C. Lindberg and Robert S. Westman, 397–436. Cambridge: Cambridge University Press, 1990.
———. "Physicians and Natural History." In *Cultures of Natural History*, edited by Nicholas Jardine, James A. Secord, and Emma C. Spary, 91–105. Cambridge: Cambridge University Press, 1996.
———. "Victories for Empiricism, Failures for Theory: Medicine and Science in the Seventeenth Century." In *The Body as Object and Instrument of Knowledge: Embodied

Empiricism in Early Modern Science, edited by Charles T. Wolfe and Ofer Gal, 9–32. Dordrecht: Springer, 2010.

Copenhaver, Brian P. "Astrology and Magic." In *The Cambridge History of Renaissance Philosophy*, edited by Charles B. Schmitt, Quentin Skinner, Eckhard Kessler, and Jill Kraye, 264–300. Cambridge: Cambridge University Press, 1988.

———. *Magic in Western Culture: From Antiquity to the Enlightenment*. Cambridge: Cambridge University Press, 2015.

———. "Natural Magic, Hermetism, and Occultism in Early Modern Science." In *Reappraisals of the Scientific Revolution*, edited by David C. Lindberg and Robert S. Westman, 261–301. Cambridge: Cambridge University Press, 1990.

———. "Renaissance Magic and Neoplatonic Philosophy: 'Ennead' 4.3–5 in Ficino's 'De vita coelitus comparanda.'" In *Marsilio Ficino e il ritorno di Platone: Studi e documenti*, edited by Gian C. Garfagnini, 2:351–69. Florence: Leo S. Olschki, 1986.

———. "Scholastic Philosophy and Renaissance Magic in the De Vita of Marsilio Ficino." *Renaissance Quarterly* 37, no. 4 (1984): 523–54.

———. "A Tale of Two Fishes: Magical Objects in Natural History from Antiquity Through the Scientific Revolution." *Journal of the History of Ideas* 52, no. 3 (1991): 373–98.

Coudert, Allison P. "Henry More, the Kabbalah, and the Quakers." In *Philosophy, Science, and Religion in England (1640–1700)*, edited by Richard Kroll, Richard Ashcraft, and Perez Zagorin, 31–67. Cambridge: Cambridge University Press, 1992.

Crisciani, Chiara. "Histories, Stories, *Exempla*, and Anecdotes: Michele Savonarola from Latin to Vernacular." In *Historia: Empiricism and Erudition in Early Modern Europe*, edited by Gianna Pomata and Nancy G. Siraisi, 297–324. Cambridge, Mass.: MIT Press, 2005.

Crocker, Robert, ed. *Religion, Reason, and Nature in Early Modern Europe*. Dordrecht: Kluwer, 2001.

Cunningham, Andrew. "Identifying Diseases in the Past: Cutting Through the Gordian Knot." *Asclepio* 54 (2002): 13–34.

Curth, Louise H. *English Almanacs, Astrology, and Popular Medicine: 1550–1700*. Manchester: Manchester University Press, 2007.

———. "The Medical Content of English Almanacs, 1640–1700." *Journal of the History of Medicine and Allied Sciences* 60, no. 3 (2005): 255–82.

Dan, Joseph, ed. *The Christian Kabbalah: Jewish Mystical Books and Their Christian Interpreters*. Cambridge, Mass.: Harvard College Library, 1997.

———. *Kabbalah: A Very Short Introduction*. Oxford: Oxford University Press, 2006.

Daston, Lorraine, and Peter Galison. *Objectivity*. New York: Zone Books, 2007.

Daston, Lorraine, and Elizabeth Lunbeck, eds. *Histories of Scientific Observation*. Chicago: University of Chicago Press, 2011.

Davies, Owen. "Angels in Elite and Popular Magic, 1650–1790." In *Angels in the Early Modern World*, edited by Peter Marshall and Alexandra Walsham, 297–319. New York: Cambridge University Press, 2006.

———. *Grimoires: A History of Magic Books*. Oxford: Oxford University Press, 2009.

Debus, Allen G. *The English Paracelsians*. London: Oldbourne Press, 1965.

———. *The French Paracelsians: The Chemical Challenge to Medical and Scientific Tradition in Early Modern France*. Cambridge: Cambridge University Press, 1991.

———. "Paracelsian Medicine: Noah Biggs and the Problem of Medical Reform." In *Medicine in Seventeenth-Century England: A Symposium Held at UCLA in Honour of C. D. O'Malley*, edited by Allen G. Debus, 33–48. Berkeley: University of California Press, 1974.

Dick, Hugh G. "Students of Physic and Astrology: A Survey of Astrological Medicine in the Age of Science." *Journal of the History of Medicine and Allied Sciences* 1 (1946): 300–15.

Duden, Barbara. *The Woman Beneath the Skin: A Doctor's Patients in Eighteenth-Century Germany*. Translated by Thomas Dunlap. Cambridge, Mass.: Harvard University Press, 1991. Originally published in 1987 in German.

Dyrness, William A. *Reformed Theology and Visual Culture: The Protestant Imagination from Calvin to Edwards.* Cambridge: Cambridge University Press, 2004.
Ebeling, Florian. *The Secret History of Hermes Trismegistus: Hermeticism from Ancient to Modern Times.* Translated by David Lorton. Ithaca: Cornell University Press, 2007.
Eire, Carlos M. N. *War Against the Idols: The Reformation of Worship from Erasmus to Calvin.* Cambridge: Cambridge University Press, 1986.
Elmer, Peter. "Medicine, Religion, and the Puritan Revolution." In *The Medical Revolution of the Seventeenth Century*, edited by Roger French and Andrew Wear, 10–45. Cambridge: Cambridge University Press, 1989.
Elton, Geoffrey R. "A High Road to Civil War?" In *From the Renaissance to the Counter-Reformation: Essays in Honour of Garrett Mattingly*, edited by Charles H. Carter, 164–82. London: Cape, 1966.
Eppley, Daniel. "Jackson, Henry (1585/6–1662)." In *Oxford Dictionary of National Biography*, edited by H. C. G. Matthew and B. Harrison, 29:485–86. New York: Oxford University Press, 2004.
Evans, Gillian R. *Problems of Authority in the Reformation Debates.* Cambridge: Cambridge University Press, 1992.
Evans, Joan. *Magical Jewels of the Middle Ages and the Renaissance Particularly in England.* Oxford: Clarendon Press, 1922.
Fanger, Claire, ed. *Conjuring Spirits: Texts and Traditions of Medieval Ritual Magic.* University Park: Penn State University Press, 1998.
———. "Introduction: Theurgy, Magic, and Mysticism." In *Invoking Angels: Theurgic Ideas and Practices, Thirteenth to Sixteenth Centuries*, edited by Claire Fanger, 1–33. University Park: Penn State University Press, 2012.
———. "Virgin Territory: Purity and Divine Knowledge in Late Medieval Catoptromantic Texts." *Aries* 5, no. 2 (2005): 200–25.
Farmer, Steve A. *Syncretism in the West: Pico's 900 Theses (1486); The Evolution of Traditional, Religious, and Philosophical Systems.* Tempe, Ariz.: Medieval and Renaissance Texts and Studies, 1998.
Ferrell, Lori A., and Peter E. McCullough, eds. *The English Sermon Revised: Religion, Literature, and History, 1600–1750.* Manchester: Manchester University Press, 2001.
Firth, Katharine R. *The Apocalyptic Tradition in Reformation Britain, 1530–1645.* Oxford Historical Monographs. Oxford: Oxford University Press, 1979.
Fissell, Mary E. "The Disappearance of the Patient's Narrative and the Invention of Hospital Medicine." In *British Medicine in an Age of Reform*, edited by Roger French and Andrew Wear, 92–109. London: Routledge, 1991.
Fletcher, Angus. *Allegory: The Theory of a Symbolic Mode.* Ithaca: Cornell University Press, 1964.
French, Peter. *John Dee: The World of an Elizabethan Magus.* London: Routledge and Kegan Paul, 1972.
French, Roger. "Astrology in Medical Practice." In *Practical Medicine from Salerno to the Black Death*, edited by Luis García-Ballester, Roger French, Jon Arrizabalaga, and Andrew Cunningham, 30–59. Cambridge: Cambridge University Press, 1994.
Gaskill, Malcolm. "Witchcraft in Early Modern Kent: Stereotypes and the Background to Accusations." In *Witchcraft in Early Modern Europe: Studies in Culture and Belief*, edited by Jonathan Barry, Marianne Hester, and Gareth Roberts, 257–87. Cambridge: Cambridge University Press, 1996.
Geneva, Ann. *Astrology and the Seventeenth-Century Mind: William Lilly and the Language of the Stars.* Manchester: Manchester University Press, 1995.
Gentilcore, David. "The Fear of Disease and the Disease of Fear." In *Fear in Early Modern Society*, edited by William G. Naphy and Penny Roberts, 184–208. Manchester: Manchester University Press, 1997.
———. *Healers and Healing in Early Modern Italy.* Manchester: Manchester University Press, 1998.

Glennie, Paul, and Nigel Thrift. *Shaping the Day: A History of Timekeeping in England and Wales, 1300–1800*. Oxford: Oxford University Press, 2009.

Goldwater, Leonard J. *Mercury: A History of Quicksilver*. Baltimore: York Press, 1972.

Gordon, Bruce. "The Renaissance Angel." In *Angels in the Early Modern World*, edited by Peter Marshall and Alexandra Walsham, 41–63. New York: Cambridge University Press, 2006.

Gow, Andrew C. *The Red Jews: Antisemitism in an Apocalyptic Age, 1200–1600*. Studies in Medieval and Reformation Thought. Leiden: E. J. Brill, 1995.

Graf, Fritz. "Augustine and Magic." In *The Metamorphosis of Magic from Late Antiquity to the Early Modern Period*, edited by Jan N. Bremmer and Jan R. Veenstra, 87–103. Leuven: Peeters, 2002.

Grafton, Anthony. *Cardano's Cosmos: The Worlds and Works of a Renaissance Astrologer*. Cambridge, Mass.: Harvard University Press, 1999.

———. *Defenders of the Text: The Traditions of Scholarship in an Age of Science, 1450–1800*. Cambridge, Mass.: Harvard University Press, 1991.

Grafton, Anthony, and Nancy G. Siraisi. "Between the Election and My Hopes: Girolamo Cardano and Medical Astrology." In *Secrets of Nature: Astrology and Alchemy in Early Modern Europe*, edited by William R. Newman and Anthony Grafton, 69–131. Cambridge, Mass.: MIT Press, 2001.

Grant, Edward. *Planets, Stars, and Orbs: The Medieval Cosmos, 1200–1687*. Cambridge: Cambridge University Press, 1994.

Green, Ian. *Print and Protestantism in Early Modern England*. Oxford: Oxford University Press, 2000.

Grell, Ole P., ed. *Paracelsus: The Man and His Reputation, His Ideas, and Their Transformation*. Leiden: Brill, 1998.

Hacking, Ian. *The Emergence of Probability: A Philosophical Study of Early Ideas About Probability, Induction, and Statistical Inference*. 2nd ed. New York: Cambridge University Press, 2006.

Haigh, Christopher. "The Character of an Antipuritan." *Sixteenth Century Journal* 35, no. 3 (2004): 671–88.

Hallacker, Anja. "On Angelic Bodies: Some Philosophical Discussions in the Seventeenth Century." In *Angels in Medieval Philosophical Inquiry: Their Function and Significance*, edited by Isabel Iribarren and Martin Lenz, 201–14. Aldershot: Ashgate, 2008.

Hanegraaff, Wouter J. *Esotericism and the Academy: Rejected Knowledge in Western Culture*. Cambridge: Cambridge University Press, 2012.

Harkness, Deborah E. *John Dee's Conversations with Angels: Cabala, Alchemy, and the End of Nature*. Cambridge: Cambridge University Press, 1999.

Harley, David. "James Hart of Northampton and the Calvinist Critique of Priest-Physicians: An Unpublished Polemic of the Early 1620s." *Medical History* 42, no. 3 (1998): 362–86.

———. "Mental Illness, Magical Medicine, and the Devil in Northern England, 1650–1700." In *The Medical Revolution of the Seventeenth Century*, edited by Roger French and Andrew Wear, 114–44. Cambridge: Cambridge University Press, 1989.

Harris, Jason M. *Folklore and the Fantastic in Nineteenth-Century British Fiction*. Aldershot: Ashgate, 2008.

Harrison, Mark. "From Medical Astrology to Medical Astronomy: Sol-Lunar and Planetary Theories of Disease in British Medicine, c. 1700–1850." *British Journal for the History of Science* 33, no. 1 (2000): 25–48.

Harrison, Peter. *The Bible, Protestantism, and the Rise of Natural Science*. Cambridge: Cambridge University Press, 1998.

———. *The Fall of Man and the Foundations of Science*. Cambridge: Cambridge University Press, 2008.

———. "Hermeneutics and Natural Knowledge in the Reformers." In *Nature and Scripture in the Abrahamic Religions, up to 1700*, edited by Jiste M. van der Meer and Scott Mandelbrote, 1:341–62. Brill's Series in Church History. Leiden: Brill, 2008.

Harvey, Kate. "The Role of Angels in English Protestant Thought, 1580 to 1660." PhD diss., Cambridge University, 2005.
Hesse, Mary. "Hermeticism and Historiography: An Apology for the Internal History of Science." In *Historical and Philosophical Perspectives of Science*, edited by Roger H. Stuewer, 134–60. Minneapolis: University of Minnesota Press, 1970.
Heyd, Michael. *Be Sober and Reasonable: The Critique of Enthusiasm in the Seventeenth and Early Eighteenth Centuries*. Brill's Studies in Intellectual History. Leiden: Brill, 1995.
Hill, Christopher. *Antichrist in Seventeenth-Century England*. 1971. Reprint, London: Oxford University Press, 1990.
———. *The English Revolution, 1640*. London: Lawrence and Wishart, 1940.
The History of the University of Oxford. Vol. 4, *Seventeenth-Century Oxford*, edited by Nicholas Tyacke. Oxford: Oxford University Press, 1997.
Huffman, William H. *Robert Fludd and the End of the Renaissance*. New York: Routledge, 1988.
Hughes, Ann. "The Meanings of Religious Polemic." In *Puritanism: Transatlantic Perspectives on a Seventeenth-Century Anglo-American Faith*, edited by Francis J. Bremer, 201–29. Boston: Massachusetts Historical Society, 1993.
Hunter, Michael. *John Aubrey and the Realm of Learning*. London: Duckworth, 1975.
Hunter, Michael, Antonio Clericuzio, and Lawrence M. Principe, eds. *The Correspondence of Robert Boyle, 1636–1691*. 6 vols. London: Pickering & Chatto, 2001.
Hutchison, Keith. "What Happened to Occult Qualities in the Scientific Revolution?" *Isis* 73 (1982): 233–53.
Hutton, Sarah. "Of Physics and Philosophy: Anne Conway, F. M. van Helmont, and Seventeenth-Century Medicine." In *Religio Medici: Religion and Medicine in Seventeenth-Century England*, edited by Ole Peter Grell and Andrew Cunningham, 228–46. Aldershot: Scolar Press, 1996.
———. "Thomas Jackson, Oxford Platonist, and William Twisse, Aristotelian." *Journal of the History of Ideas* 39, no. 4 (1978): 635–52.
Iribarren, Isabel, and Martin Lenz, eds. *Angels in Medieval Philosophical Inquiry: Their Function and Significance*. Aldershot: Ashgate, 2008.
Jeanneret, Michel. "Renaissance Exegesis." In *The Cambridge History of Literary Criticism*, vol. 3, *The Renaissance*, edited by Glyn P. Norton, 36–43. Cambridge: Cambridge University Press, 1999.
Jenkins, Allan K., and Patrick Preston. *Biblical Scholarship and the Church: A Sixteenth-Century Crisis of Authority*. Ashgate New Critical Thinking in Religion, Theology and Biblical Studies. Aldershot: Ashgate, 2007.
Jenner, Mark S. R., and Patrick Wallis, eds. *Medicine and the Market in England and Its Colonies, c. 1450–c. 1850*. Basingstoke: Palgrave Macmillan, 2007.
Kassell, Lauren. "The Economy of Magic in Early Modern England." In *The Practice of Reform in Health, Medicine, and Science, 1500–2000: Essays for Charles Webster*, edited by Margaret Pelling and Scott Mandelbrote, 43–57. Aldershot: Ashgate, 2005.
———. *Medicine and Magic in Elizabethan London: Simon Forman; Astrologer, Alchemist, and Physician*. Oxford Historical Monographs. Oxford: Clarendon Press, 2005.
Keck, David. *Angels and Angelology in the Middle Ages*. New York: Oxford University Press, 1998.
Kelly, John T. *Practical Astronomy During the Seventeenth Century: Almanac-Makers in America and England*. New York: Garland, 1991.
Kelsey, Sean. "Booth, George, First Baron Delamer [Delamere] (1622–1684)." In *Oxford Dictionary of National Biography*, edited by H. C. G. Matthew and B. Harrison, 6:613–15. New York: Oxford University Press, 2004.
Kieckhefer, Richard. "Did Magic Have a Renaissance? An Historiographic Question Revisited." In *Magic and the Classical Tradition*, edited by Charles Burnett and W. F. Ryan, 199–212. London: Warburg Institute; Turin: Nino Aragno Editore, 2006.

---. *Forbidden Rites: A Necromancer's Manual of the Fifteenth Century*. 1997. Reprint, University Park: Penn State University Press, 2003.

---. *Magic in the Middle Ages*. 1989. Reprint, Cambridge: Cambridge University Press, 1991.

---. "The Specific Rationality of Medieval Magic." *American Historical Review* 99, no. 3 (1994): 813–36.

Kiessling, Nicholas K., ed. *The Life of Anthony Wood in His Own Words*. Oxford: Bodleian Library, 2009.

Killeen, Kevin. *Biblical Scholarship, Science, and Politics in Early Modern England: Thomas Browne and the Thorny Place of Knowledge*. Surrey: Ashgate, 2009.

Killeen, Kevin, and Peter J. Forshaw, eds. *The Word and the World: Biblical Exegesis and Early Modern Science*. New York: Palgrave Macmillan, 2007.

King, Lester. "What Is Disease?" In *Concepts of Health and Disease: Interdisciplinary Perspectives*, edited by Arthur L. Caplan, H. Tristram Engelhardt Jr., and James J. McCartney, 107–18. Reading, Mass.: Addison-Wesley, 1981.

Klaassen, Frank. "English Manuscripts of Magic, 1300–1500: A Preliminary Survey." In *Conjuring Spirits: Texts and Traditions of Medieval Ritual Magic*, edited by Claire Fanger, 3–31. University Park: Penn State University Press, 1998.

---. "Medieval Ritual Magic in the Renaissance." *Aries: Journal of the Study of Western Esotericism* 3, no. 2 (2003): 166–99.

---. "Ritual Invocation and Early Modern Science: The Skrying Experiments of Humphrey Gilbert." In *Invoking Angels: Theurgic Ideas and Practices, Thirteenth to Sixteenth Centuries*, edited by Claire Fanger, 341–66. University Park: Penn State University Press, 1998.

---. *The Transformations of Magic: Illicit Learned Magic in the Later Middle Ages and Renaissance*. University Park: Penn State University Press, 2012.

Kuriyama, Shigehisa. "The Forgotten Fear of Excrement." *Journal of Medieval and Early Modern Studies* 38, no. 3 (2008): 413–42.

---. "Interpreting the History of Bloodletting." *Journal of the History of Medicine and Allied Sciences* 50, no. 1 (1995): 11–46.

Lake, Peter. *Anglicans and Puritans? Presbyterianism and English Conformist Thought from Whitgift to Hooker*. London: Unwin Hyman, 1988.

---. "Anti-Puritanism: The Structure of a Prejudice." In *Religious Politics in Post-Reformation England: Essays in Honour of Nicholas Tyacke*, edited by Kenneth Fincham and Peter Lake, 80–97. Woodbridge: Boydell Press, 2006.

---. "Defining Puritanism: Again?" In *Puritanism: Transatlantic Perspectives on a Seventeenth-Century Anglo-American Faith*, edited by Francis J. Bremer, 3–29. Boston: Massachusetts Historical Society, 1993.

---. "Joseph Hall, Robert Skinner, and the Rhetoric of Moderation at the Early Stuart Court." In *The English Sermon Revised: Religion, Literature, and History, 1600–1750*, edited by Lori A. Ferrell and Peter E. McCullough, 167–85. Manchester: Manchester University Press, 2001.

---. "Puritanism, (Monarchical) Republicanism, and Monarchy: Or, John Whitgift, Antipuritanism, and the 'Invention' of Popularity." *Journal of Medieval and Early Modern Studies* 40 (2010): 463–95.

Lake, Peter, and Michael Questier. *The Antichrist's Lewd Hat: Protestants, Papists, and Players in Post-Reformation England*. New Haven: Yale University Press, 2002.

Landes, David S. *Revolution in Time: Clocks and the Making of the Modern World*. Cambridge, Mass.: Harvard University Press, 1983.

Láng, Benedek. "Angels Around the Crystal: The Prayer Book of King Wladislas and the Treasure Hunts of Henry the Bohemian." *Aries: Journal of the Study of Western Esotericism* 5, no. 1 (2005): 1–32.

Lidaka, Juris G. "The Book of Angels, Rings, Characters, and Images of the Planets: Attributed to Osbern Bokenham." In *Conjuring Spirits: Texts and Traditions of Medieval*

Ritual Magic, edited by Claire Fanger, 32–75. University Park: Penn State University Press, 1998.
Louth, Andrew. "Pagan Theurgy and Christian Sacramentalism in Denys the Areopagite." *Journal of Theological Studies* 37, no. 2 (1986): 432–38.
MacDonald, Michael. "The Career of Astrological Medicine in England." In *Religio Medici: Religion and Medicine in Seventeenth-Century England*, edited by Ole P. Grell and Andrew Cunningham, 62–90. Aldershot: Scolar Press, 1996.
———. "An Early Seventeenth-Century Defence of Usury." *Historical Research* 60 (1987): 353–60.
———. *Mystical Bedlam: Madness, Anxiety, and Healing in Seventeenth-Century England*. Cambridge: Cambridge University Press, 1981.
MacDonald Ross, G. "Occultism and Philosophy in the Seventeenth Century." In *Philosophy, Its History and Historiography*, edited by Alan J. Holland, 95–115. Dordrecht: D. Reidel, 1985.
Macfarlane, Alan. *Witchcraft in Tudor and Stuart England: A Regional and Comparative Study*. New York: Harper & Row, 1970.
Maclean, Ian. *Logic, Signs, and Nature in the Renaissance: The Case of Learned Medicine*. Cambridge: Cambridge University Press, 2002.
———. "The Science of Nature and the Science of God: Conflict and Collaboration in the Early Modern Period." *Filozofia* 63 (2008): 352–64.
Malcolm, Noel. *Aspects of Hobbes*. Oxford: Clarendon Press, 2002.
Marshall, Peter. "Angels Around the Deathbed: Variations on a Theme in the English Art of Dying." In *Angels in the Early Modern World*, edited by Peter Marshall and Alexandra Walsham, 83–103. New York: Cambridge University Press, 2006.
Marshall, Peter, and Alexandra Walsham, eds. *Angels in the Early Modern World*. New York: Cambridge University Press, 2006.
———. "Migrations of Angels in the Early Modern World." In *Angels in the Early Modern World*, edited by Peter Marshall and Alexandra Walsham, 1–40. New York: Cambridge University Press, 2006.
Martin, David. *On Secularization: Towards a Revised General Theory*. Aldershot: Ashgate, 2005.
Mayr-Harting, Henry. *Perceptions of Angels in History: An Inaugural Lecture Delivered in the University of Oxford on 14 November 1997*. Oxford: Clarendon Press, 1998.
McGinn, Bernard. *Antichrist: Two Thousand Years of the Human Fascination with Evil*. New York: HarperCollins, 1994.
McGrath, Alister E. *The Intellectual Origins of the European Reformation*. New York: Blackwell, 1987.
McKim, Donald K., ed. *Calvin and the Bible*. Cambridge: Cambridge University Press, 2006.
Meier-Oeser, Stephan. "Medieval, Renaissance, and Reformation Angels: A Comparison." In *Angels in Medieval Philosophical Inquiry: Their Function and Significance*, edited by Isabel Iribarren and Martin Lenz, 187–200. Aldershot: Ashgate, 2008.
Merkel, Ingrid, and Allen G. Debus, eds. *Hermeticism and the Renaissance: Intellectual History and the Occult in Early Modern Europe*. Washington: Folger Books, 1988.
Merton, Robert K. "Science, Technology, and Society in Seventeenth-Century England." *Osiris* 4, no. 2 (1938): 360–632.
Milton, Anthony. *Catholic and Reformed: The Roman and Protestant Churches in English Protestant Thought, 1600–1640*. Cambridge: Cambridge University Press, 1995.
Mohamed, Feisal G. *In the Anteroom of Divinity: The Reformation of Angels from Colet to Milton*. Toronto: University of Toronto Press, 2008.
Monod, Paul Kléber. *Solomon's Secret Arts: The Occult in the Age of Enlightenment*. New Haven: Yale University Press, 2013.
Nance, Brian. *Turquet de Mayerne as Baroque Physician: The Art of Medical Portraiture*. Clio Medica 65, Wellcome Series in the History of Medicine. Amsterdam: Rodopi, 2001.

New, John G. H. *Anglican and Puritan: The Basis of Their Opposition, 1558–1640*. Stanford: Stanford University Press, 1964.
Niccoli, Ottavia. *Prophecy and People in Renaissance Italy*. Princeton: Princeton University Press, 1990.
North, John D. *Horoscopes and History*. London: Warburg Institute, University of London, 1986.
Nothaft, Carl P. E. *Dating the Passion: The Life of Jesus and the Emergence of Scientific Chronology (200–1600)*. Leiden: Brill, 2012.
———. "From Sukkot to Saturnalia: The Attack on Christmas in Sixteenth-Century Chronological Scholarship." *Journal of the History of Ideas* 72, no. 4 (2011): 503–22.
Nutton, Vivian. "The Fortunes of Galen." In *The Cambridge Companion to Galen*, edited by R. J. Hankinson, 355–89. New York: Cambridge University Press, 2008.
O'Neil, Mary R. "'*Sacerdote ovvero strione*': Ecclesiastical and Superstitions Remedies in Sixteenth-Century Italy." In *Understanding Popular Culture: Europe from the Middle Ages to the Nineteenth Century*, edited by Steven L. Kaplan, 53–83. Berlin: Mouton, 1984.
Page, Sophie. "Image-Magic Texts and a Platonic Cosmology at St Augustine's, Canterbury, in the Late Middle Ages." In *Magic and the Classical Tradition*, edited by William Francis Ryan and Charles S. F. Burnett, 69–98. London: Warburg Institute; Turin: Nino Aragno Editore, 2006.
———. *Magic in the Cloister: Pious Motives, Illicit Interests, and Occult Approaches to the Medieval Universe*. University Park: Penn State University Press, 2013.
———. *Magic in Medieval Manuscripts*. London: British Library, 2004.Park, Katharine. "Medicine and Magic: The Healing Arts." In *Gender and Society in Renaissance Europe*, edited by Judith C. Brown and Robert C. Davis, 129–49. London: Addison Wesley Longman, 1998.
Parry, Glyn. *The Arch-Conjurer of England: John Dee*. New Haven: Yale University Press, 2011.
Parry, Graham. "Wood, Anthony [Anthony à Wood] (1632–1695)." In *Oxford Dictionary of National Biography*, edited by H. C. G. Matthew and B. Harrison, 60:67–70. New York: Oxford University Press, 2004.
Partee, Charles. *The Theology of John Calvin*. Louisville: Westminster John Knox, 2008.
Pelling, Margaret. *The Common Lot: Sickness, Medical Occupations, and the Urban Poor in Early Modern England*. London: Longman, 1998.
———. "Defensive Tactics: Networking by Female Medical Practitioners in Early Modern London." In *Communities in Early Modern England: Networks, Place, Rhetoric*, edited by Alexandra Shepard and Phil Withington, 38–53. Manchester: Manchester University Press, 2000.
———. "Knowledge Common and Acquired: The Education of Unlicensed Medical Practitioners in Early Modern London." In *History of Medical Education in Britain*, edited by Vivian Nutton and Roy Porter, 250–79. Amsterdam: Rodopi, 1995.
———. "Public and Private Dilemmas: The College of Physicians in Early Modern London." In *Medicine, Health, and the Public Sphere in Britain, 1600–2000*, edited by Steve Sturdy, 27–42. London: Routledge, 2002.
Pelling, Margaret, and Charles Webster. "Medical Practitioners." In *Health, Medicine, and Mortality in the Sixteenth Century*, edited by Charles Webster, 165–235. Cambridge: Cambridge University Press, 1979.
Pelling, Margaret, and Frances White. *Medical Conflicts in Early Modern London: Patronage, Physicians, and Irregular Practitioners, 1550–1640*. Oxford Studies in Social History. Oxford: Clarendon Press, 2003.
Perrone Compagni, Vittoria. "'Dispersa Intentio': Alchemy, Magic, and Scepticism in Agrippa." *Early Science and Medicine* 5, no. 2 (2000): 160–77.
Peterson, Joseph H. *The Clavis or Key to the Magic of Solomon: From an Original Talismanic Grimoire in Full*. Lake Worth, Fla.: Ibis Press, 2009.

Pickstone, John V. *Ways of Knowing: A New History of Science, Technology, and Medicine*. Chicago: University of Chicago Press, 2001.
Pingree, David. "The Diffusion of Arabic Magical Texts in Western Europe." In *La diffusione delle scienze islamiche nel medio evo europeo*, edited by Biancamaria Scarcia Amoretti, 57–102. Rome: Accademia Nazionale di Lincei, 1987.
———. "Learned Magic in the Time of Frederick II." In *Le scienze alla corte di Federico II / Sciences at the Court of Frederick II*, 39–56. Micrologus Library 2. Turnhout: Brepols, 1994.
Pomata, Gianna. *Contracting a Cure: Patients, Healers, and the Law in Early Modern Bologna*. Baltimore: Johns Hopkins University Press, 1998.
———. "Menstruating Men: Similarity and Difference of the Sexes in Early Modern Medicine." In *Generation and Degeneration: Tropes of Reproduction in Literature and History from Antiquity Through Early Modern Europe*, edited by Valeria Finucci and Kevin Brownlee, 109–52. Durham: Duke University Press, 2001.
———. "*Praxis Historialis*: The Uses of Historia in Early Modern Medicine." In *Historia: Empiricism and Erudition in Early Modern Europe*, edited by Gianna Pomata and Nancy G. Siraisi, 105–46. Cambridge, Mass.: MIT Press, 2005.
Poole, William. *John Aubrey and the Advancement of Learning*. Oxford: Bodleian Library, 2010.
Porter, Roy. *Health for Sale: Quackery in England, 1660–1850*. Manchester: Manchester University Press, 1989.
———. "The Patient's View: Doing Medical History from Below." *Theory and Society* 14, no. 2 (1985): 175–98.
Press, Gerald A. "The Subject and Structure of Augustine's *De Doctrina Christiana*." *Augustinian Studies* 11 (1980): 99–124.
Principe, Lawrence M. *The Aspiring Adept: Robert Boyle and His Alchemical Quest*. Princeton: Princeton University Press, 1998.
Prior, Charles. *Defining the Jacobean Church: The Politics of Religious Controversy, 1603–1625*. Cambridge: Cambridge University Press, 2005.
Quantin, Jean-Louis. *The Church of England and Christian Antiquity: The Construction of Confessional Identity in the Seventeenth Century*. Oxford: Oxford University Press, 2009.
Rankin, Alisha M. "Duchess, Heal Thyself: Elisabeth of Rochlitz and the Patient's Perspective in Early Modern Germany." *Bulletin of the History of Medicine* 82, no. 1 (2008): 109–44.
———. *Panaceia's Daughters: Noblewomen as Healers in Early Modern Germany*. Chicago: University of Chicago Press, 2013.
Raymond, Joad, ed. *Conversations with Angels: Essays Towards a History of Spiritual Communication, 1100–1700*. Basingstoke: Palgrave Macmillan, 2011.
———. *Milton's Angels: The Early Modern Imagination*. Oxford: Oxford University Press, 2010.
———. "With the Tongues of Angels: Angelic Conversations in *Paradise Lost* and Seventeenth-Century England." In *Angels in the Early Modern World*, edited by Peter Marshall and Alexandra Walsham, 256–81. New York: Cambridge University Press, 2006.
Renaker, David, "The Horoscope of Christ." *Milton Studies* 12 (1978): 213–33.
Riley, James C. *Rising Life Expectancy: A Global History*. Cambridge: Cambridge University Press, 2001.
Risse, Guenter B., and John H. Warner. "Reconstructing Clinical Activities: Patient Records in Medical History." *Social History of Medicine* 5, no. 2 (1992): 183–205.
Roos, Anna M. "Luminaries in Medicine: Richard Mead, James Gibbs, and Solar and Lunar Effects on the Human Body in Early Modern England." *Bulletin of the History of Medicine* 74, no. 3 (2000): 433–57.

———. "'Magic Coins' and 'Magic Squares': The Discovery of Astrological Sigils in the Oldenburg Letters." *Notes and Records of the Royal Society of London* 62 (2008): 271–88.
Rosen, Edward. "Was Copernicus a Hermetist?" In *Historical and Philosophical Perspectives of Science*, edited by Roger H. Stuewer, 163–71. Minneapolis: University of Minnesota Press, 1970.
Rowse, Alfred L. *Sex and Society in Shakespeare's Age: Simon Forman the Astrologer.* New York: Charles Scribner's Sons, 1974.
Russell, Conrad, ed. *The Origins of the English Civil War.* Basingstoke: Macmillan, 1973.
Rutkin, H. Darrel. "Astrology." In *The Cambridge History of Science*, vol. 3, *Early Modern Science*, edited by Katharine Park and Lorraine Daston, 541–61. Cambridge: Cambridge University Press, 2008.
Saif, Liana. *The Arabic Influences on Early Modern Occult Philosophy.* Palgrave Historical Studies in Witchcraft and Magic. London: Palgrave Macmillan, 2015.
Sangha, Laura. *Angels and Belief in England, 1480–1700.* London: Pickering & Chatto, 2012.
Sawyer, Ronald C. "Patients, Healers, and Disease in the Southeast Midlands, 1597–1634." PhD diss., University of Wisconsin–Madison, 1986.
———. "'Strangely Handled in All her Lyms': Witchcraft and Healing in Jacobean England." *Journal of Social History* 22, no. 3 (1988–89): 461–85.
Scribner, Robert W. "The Reformation, Popular Magic, and the 'Disenchantment of the World.'" *Journal of Interdisciplinary History* 23, no. 3 (1993): 475–94.
Seznec, Jean. *The Survival of the Pagan Gods: The Mythological Tradition and Its Place in Renaissance Humanism and Art.* Translated by Barbara F. Sessions. New York: Harper & Row, 1961. First published in English in 1953 by Pantheon Books.
Shapin, Steven. *The Scientific Revolution.* Chicago: University of Chicago Press, 1996.
———. *A Social History of Truth: Civility and Science in Seventeenth-Century England.* Chicago: University of Chicago Press, 1994.
Shapin, Steven, and Simon Schaffer. *Leviathan and the Air-Pump: Hobbes, Boyle, and the Experimental Life.* Princeton: Princeton University Press, 1985.
Shapiro, Barbara J. *A Culture of Fact: England, 1550–1720.* Ithaca: Cornell University Press, 2000.
———. *Probability and Certainty in Seventeenth-Century England: A Study of the Relationships Between Natural Science, Religion, History, Law, and Literature.* Princeton: Princeton University Press, 1983.
Sharpe, James A. *Witchcraft in Seventeenth-Century Yorkshire: Accusations and Counter Measures.* York: Borthwick Institute of Historical Research, 1992.
Shatzmiller, Joseph. "In Search of the 'Book of Figures': Medicine and Astrology in Montpellier at the Turn of the Fourteenth Century." *Association of Jewish Studies Review* 7/8 (1982/3): 383–407.
Siraisi, Nancy G. *The Clock and the Mirror: Girolamo Cardano and Renaissance Medicine.* Princeton: Princeton University Press, 1997.
———. "Disease and Symptom as Problematic Concepts in Renaissance Medicine." In *Res et Verba in the Renaissance*, edited by Eckhard Kessler and Ian Maclean, 217–40. Wiesbaden: Harrassowitz, 2002.
———. *Medieval and Early Renaissance Medicine: An Introduction to Knowledge and Practice.* Chicago: University of Chicago Press, 1990.
Skinner, Stephen. *Terrestrial Astrology: Divination by Geomancy.* London: Routledge and Kegan Paul, 1980.
Smoller, Laura A. "Astrology and the Sibyls: John of Legnano's *De Adventu Christi* and the Natural Theology of the Later Middle Ages." *Science in Context* 20, no. 3 (2007): 423–50.
———. *History, Prophecy, and the Stars: The Christian Astrology of Pierre d'Ailly, 1350–1420.* Princeton: Princeton University Press, 1994.

Soergel, Philip M. "Luther on the Angels." In *Angels in the Early Modern World*, edited by Peter Marshall and Alexandra Walsham, 64–82. New York: Cambridge University Press, 2006.
Spence, Lewis. *An Encyclopaedia of Occultism: A Compendium of Information on the Occult Sciences, Occult Personalities, Psychic Science, Magic, Demonology, Spiritism, and Mysticism.* New York: Dover, 2003. First published in 1920 by University Books.
Stein, Claudia. "The Meaning of Signs: Diagnosing the French Pox in Early Modern Augsburg." *Bulletin of the History of Medicine* 80, no. 4 (2006): 617–48.
———. *Negotiating the French Pox in Early Modern Germany*. Aldershot: Ashgate, 2009.
Stephenson, Marcia. "From Marvelous Antidote to the Poison of Idolatry: The Transatlantic Role of Andean Bezoar Stones During the Late Sixteenth and Early Seventeenth Centuries." *Hispanic American Historical Review* 90, no. 1 (2010): 3–39.
Stolberg, Michael. "The Decline of Uroscopy in Early Modern Learned Medicine (1500–1650)." *Early Science and Medicine* 12 (2007): 313–36.
———. *Experiencing Illness and the Sick Body in Early Modern Culture*. Basingstoke: Palgrave Macmillan, 2011.
Stroumsa, Guy G. *A New Science: The Discovery of Religion in the Age of Reason*. Cambridge, Mass.: Harvard University Press, 2010.
Strype, John. *The Life and Acts of John Whitgift D.D.* Vol. 1 of 4. Oxford: Clarendon Press, 1822.
Temkin, Owsei. *Galenism: Rise and Decline of a Medical Philosophy*. Ithaca: Cornell University Press, 1973.
Thomas, Keith. *Religion and the Decline of Magic: Studies in Popular Beliefs in Sixteenth- and Seventeenth-Century England*. London: Weidenfeld & Nicolson, 1971.
Thompson, John L. "Calvin as Biblical Interpreter." In *The Cambridge Companion to John Calvin*, edited by Donald K. McKim, 58–73. Cambridge: Cambridge University Press, 2004.
Thorndike, Lynn. *A History of Magic and Experimental Science*. 8 vols. New York: Columbia University Press, 1923–41.
———. "Traditional Medieval Tracts Concerning Engraved Astrological Images." In *Mélanges Auguste Pelzer*, 217–74. Leuven: Bibliothèque de l'Université, 1947.
Torrance Kirby, William J. *Richard Hooker, Reformer and Platonist*. Aldershot: Ashgate, 2005.
Torrey, Edwin F., and Judy Miller. *The Invisible Plague: The Rise of Mental Illness from 1750 to the Present*. New Brunswick: Rutgers University Press, 2001.
Traister, Barbara H. *The Notorious Astrological Physician of London: Works and Days of Simon Forman*. Chicago: University of Chicago Press, 2001.
Trevor-Roper, Hugh R. *Europe's Physician: The Various Life of Sir Theodore de Mayerne*. New Haven: Yale University Press, 2006.
van der Meer, Jitse M., and Scott Mandelbrote, eds. *Nature and Scripture in the Abrahamic Religions, up to 1700*. 2 vols. Brill's Series in Church History. Leiden: Brill, 2008.
van Leeuwen, Henry G. *The Problem of Certainty in English Thought, 1630–1690*. The Hague: Martinus Nijhoff, 1963.
Véronèse, Julien. "Magic, Theurgy, and Spirituality in the Medieval Ritual of the *Ars notoria*." Translated by Claire Fanger. In *Invoking Angels: Theurgic Ideas and Practices, Thirteenth to Sixteenth Centuries*, edited by Claire Fanger, 37–78. University Park: Penn State University Press, 2012.
Vickers, Brian. "Analogy Versus Identity: The Rejection of Occult Symbolism, 1580–1680." In *Occult and Scientific Mentalities in the Renaissance*, edited by Brian Vickers, 95–163. Cambridge: Cambridge University Press, 1984.
———. "Frances Yates and the Writing of History." *Journal of Modern History* 51, no. 2 (1979): 287–316.

Walsham, Alexandra. "Angels and Idols in England's Long Reformation." In *Angels in the Early Modern World*, edited by Peter Marshall and Alexandra Walsham, 134–67. New York: Cambridge University Press, 2006.
———. "Catholic Reformation and the Cult of Angels in Early Modern England." In *Conversations with Angels: Essays Towards a History of Spiritual Communication, 1100–1700*, edited by Joad Raymond, 273–94. Basingstoke: Palgrave Macmillan, 2011.
———. *Catholic Reformation in Protestant Britain*. Ashgate: Farnham, 2014.
———. "Historiographical Reviews: The Reformation and 'The Disenchantment of the World' Reassessed." *Historical Journal* 51, no. 2 (2008): 497–528.
———. "Impolitic Pictures: Providence, History, and the Iconography of Protestant Nationhood in Early Stuart England." In *The Church Retrospective*, edited by Robert N. Swanson, 307–28. Studies in Church History 33. Woodbridge: Boydell Press, 1997.
———. "Invisible Helpers: Angelic Intervention in Post-Reformation England." *Past and Present* 208, no. 1 (2010): 77–130.
Warner, John H. "The Uses of Patient Records by Historians: Patterns, Possibilities, and Perplexities." *Health and History* 1 (1999): 101–11.
Watt, Tessa. *Cheap Print and Popular Piety, 1550–1640*. Cambridge: Cambridge University Press, 1991.
Wear, Andrew. "Galen in the Renaissance." In *Galen: Problems and Prospects*, edited by Vivian Nutton, 229–62. London: Wellcome Institute for the History of Medicine, 1981.
———. *Health and Healing in Early Modern England: Studies in Social and Intellectual History*. Variorum Collected Studies Series. Aldershot: Ashgate, 1998.
———. *Knowledge and Practice in English Medicine, 1550–1680*. Cambridge: Cambridge University Press, 2000.
———. "Religious Beliefs and Medicine in Early Modern England." In *The Task of Healing: Medicine, Religion, and Gender in England and the Netherlands 1450–1800*, edited by Hilary Marland and Margaret Pelling, 145–69. Rotterdam: Erasmus, 1996.
Weber, Max. *The Protestant Ethic and the Spirit of Capitalism*. Translated by Talcott Parsons. New York: Charles Scribner's Sons, 1958. Originally published in 1904–5 in German.
———. "Science as a Vocation." In *From Max Weber: Essays in Sociology*, edited and translated by Hans H. Gerth and Charles Wright Mills, 129–56. New York: Oxford University Press, 1946.
Webster, Charles. *From Paracelsus to Newton: Magic and the Making of Modern Science*. Cambridge: Cambridge University Press, 1982.
———. *The Great Instauration: Science, Medicine, and Reform, 1626–1660*. London: Duckworth, 1975.
Weill-Parot, Nicolas. "Astral Magic and Intellectual Changes (Twelfth–Fifteenth Centuries): 'Astrological Images' and the Concept of 'Addressative' Magic." In *The Metamorphosis of Magic from Late Antiquity to the Early Modern Period*, edited by Jan N. Bremmer and Jan R. Veenstra, 167–88. Leuven: Peeters, 2002.
———. "Contriving Classical References for Talismanic Magic in the Middle Ages and the Early Renaissance." In *Magic and the Classical Tradition*, edited by Charles S. F. Burnett and William F. Ryan, 163–76. London: Warburg Institute; Turin: Nino Aragno Editore, 2006.
Weisser, Olivia. "Boils, Pushes, and Wheals: Reading Bumps on the Body in Early Modern England." *Social History of Medicine* 22, no. 2 (2009): 321–39.
Whitby, Christopher. *John Dee's Actions with Spirits*. 2 vols. New York: Garland, 1988.
White, Peter. *Predestination, Policy, and Polemic: Conflict and Consensus in the English Church from the Reformation to the Civil War*. Cambridge: Cambridge University Press, 1992.
Williams, Arnold. *The Common Expositor: An Account of the Commentaries on Genesis, 1527–1633*. Chapel Hill: University of North Carolina Press, 1948.

Wilson, Adrian. "On the History of Disease Concepts: The Case of Pleurisy." *History of Science* 38, no. 3 (2000): 271–319.
Woolf, Daniel. *The Social Circulation of the Past: English Historical Culture, 1500–1730.* Oxford: Oxford University Press, 2003.
Wrigley, Edward A., and Roger S. Schofield. *The Population History of England, 1541–1871.* London: Edward Arnold, 1981.
Yates, Frances. *Giordano Bruno and the Hermetic Tradition.* London: Routledge and Kegan Paul, 1964.
———. "The Hermetic Tradition in Renaissance Science." In *Art, Science, and History in the Renaissance*, edited by Charles S. Singleton, 255–74. Baltimore: Johns Hopkins University Press, 1967.
———. *The Rosicrucian Enlightenment.* London: Routledge and Kegan Paul, 1972.
Zambelli, Paola. *The Speculum Astronomiae and Its Enigma: Astrology, Theology, and Science in Albertus Magnus and His Contemporaries.* Dordrecht: Kluwer Academic, 1992.
Zimmels, Hirsch J. *Magicians, Theologians, and Doctors.* London: Edward Goldston and Son, 1952.

INDEX

Agrippa, Henry Cornelius, 64, 65, 110, 111, 113
Albumasar, 137, 138
alchemy, 3, 9, 95, 108, 118–19
allegory, 133–35
almanacs, 25, 29
amulets, 63, 68, 77, 96. *See also* sigils
Andrews, Sir William, 41n176, 99–100, 104, 105
angelomancy. *See* angels, conversing with
angels
  belief in, 94, 120
  conversing with
    historiography of, 9, 86–95, 105
    history and philosophy of, 109–10, 113
    invocation and summoning, 65, 89, 98, 109–15, 119–20
    as a means of diagnosis, 102–3
    as a means of divination, 103–9, 121–22, 141
    in medicine, 96–104
    practitioners of, 9, 87, 95, 113–15, 121–22. *See also* John Dee; Simon Forman; Richard Napier
    using a crystal, 109, 114–15
    using a human medium, 109, 113–14, 119
  history and philosophy of, 115–16, 119–22, 125
  Raphael, the archangel, 2, 12, 62, 86, 88, 91–118 *passim*, 121
"Anglicanism." *See* English Church
Antichrist, 116–17
antimony, 35, 51, 76
Aquinas, Thomas, 63, 68, 70, 81–82, 115
Ashmole, Elias
  archive of, 5–6, 38n164, 86–88, 113, 115
  acquaintance with Napier's practice, 1n1, 5–6, 61–62, 66, 72–80, 86–88, 91, 94
  astrological and divinatory practice, 65, 72, 98n67, 105n130, 106n141, 108–10
Ashmolean Museum, Oxford, 6, 87
astral magic. *See* sigils
astrologer physicians. *See* astrological medicine, practitioners of
astrological medicine
  astrological cause, 38–39, 45–46, 48, 55

astrological tables. See *ephemeris*; *table of houses*
  handbooks and guidelines for, 15, 20, 24–25, 33, 35, 38, 51. *See also* Simon Forman, guidelines for astrological medicine
  history and philosophy of, 8–9, 14, 24–25
  interpreting the astrological chart, 37–40, 43–44
  practitioners of, 8–9. *See also* Simon Forman; Richard Napier
  signifiers of disease, 29, 31, 43–44
  validity and reliability of, 57–59
astrological talismans. *See* sigils
astrology. *See also* astrological medicine
  celestial influence, 3, 35, 39, 50, 63–83 *passim*, 110, 135
  constellation at birth, 23, 25. *See also* Christ, nativity
  erecting an astrological chart (figure), 28–31, 75
  historiography of, 54
  practitioners of, 54, 87, 110, 121, 137–38
  and religion, 59, 137–39
  role in medicine of, 2–3, 15, 53–59
  temperaments, 25–26, 44
  types of, 27–28
  as a worldview, 8, 54
Aubrey, John, 61, 86–94, 114
Augustine, 81, 125, 134, 136–37, 141
Avicenna, 68

Babington, Gervase, 132
Bacon, Roger, 137
Bacon, Sir Francis, 119n252, 142
beryl. *See* angels, conversing with, using a crystal
Bible
  biblical exegesis, 129n42, 133–35, 138–39
  interpretation of, 124–30
  references to, 83–84, 132–37
Bliss, Philip, 93
bloodletting, 1, 17n14, 19, 20, 45, 46, 77, 99–100, 102, 118. *See also* leeches

Bodleian Library, Oxford, 6
body
　early modern perception of, 15–19
　excretions, 18–20, 57
　orifices, 18–19, 46, 50
　movement of organs, 17–18
　skin, 20, 32, 46
Booker, John, 26n83, 86, 116n226
Booth, Sir George, 87, 89, 91, 93
British Library, London, 6
Bull, Frederick, 130
Burton, Robert, 47, 50, 51, 133n70

Calvin, John, 83, 116, 135
Calvinism and pro-Calvinism, 7n41, 10, 81, 123–25, 127, 132
Cardano, Girolamo, 138n103
Casaubon, Meric, 111, 118–19
Catholicism, Roman, 124–28
Chaldeans, 72, 137, 166n129
Christ, 113, 116–17, 120, 124, 126, 134
　nativity, 135–39
Chrysostom, 136
church councils, 125–27
church fathers, 58, 81, 88, 124–37 *passim*
church of St. Andrew's, Great Linford, 4
College of Physicians of London, 2, 24
Collinges, John, 138
continuity of treatment, 51–53, 102n94
"crystal ball." *See* angels, conversing with, using a crystal
Culpeper, Nicholas, 46, 48, 114n205, 116n226

Dee, Arthur, 72n68, 95n53, 114
Dee, John, 68–69, 86, 95, 113
　conversation with angels, 9, 72n68, 87, 92, 94, 95, 111, 114, 118–19
demons, 111, 113, 120
devil, 82, 83, 102, 112–13, 116, 134
diagnosis, 23–25, 32, 37–41, 43–45, 54–55, 102–3. *See also doctrine of signs;* astrological medicine
disease
　affected body parts, 35–37, 39, 46
　categories of, 23, 40
　identifying the nature of. *See* diagnosis; astrological medicine, signifiers of disease
　narrative of, 23, 33–34, 37, 52, 55–57. *See also* patients, complaints and sensations
　setting and movement of, 16, 21–22, 25, 34–36
　taxonomy of, 25, 42–45
　theory of, 17–18, 41–45. *See also* humoral medicine, history and philosophy of
diseases. *See* medical problems
"disenchantment of the world," 9–10

divination, 5, 11, 105–6. *See also* angels, conversing with; astrology
divine revelation. *See* knowledge, divine
doctrine of signs, 21, 54–55
Duden, Barbara, 17, 20, 23–24
Dunton, John, 92

emanation (of divine knowledge). *See* knowledge, divine; Neoplatonism
emetics, 1, 20, 35, 46, 50, 52, 98, 104, 106
emotions, 23–24, 43
English Church
　Articles of Faith, 125, 127, 128n37
　historiography of, 10–11
　history of, 124–28
ephemeris, 25, 29, 31
Evington, Sir James, 107–8
Exeter College, 5

Ficino, Marsilio, 46n208, 54, 64–65, 68–70, 72, 82, 110, 133
Fludd, Robert, 68n46
follow-up. *See* continuity of treatment
food, 19, 23, 33, 68
Forman, Simon
　acquaintance with Napier, 5, 66–67
　astrological and medical practice, 26, 36, 47, 72, 75–76, 106n135, 106n141, 111, 114
　guidelines for astrological medicine, 14–59 *passim*, 65–66, 79
　other papers and writings, 54, 65, 109, 111
Fox, (Doctor), 1

Galen and Galenism, 11, 16–24, 67–68, 70. *See also* humoral medicine
God
　appealing to, 111–13
　knowledge of, 84, 116, 123, 141
Great Linford. *See* Richard Napier, residence

Hall, John, 26n83, 34, 53
healing. *See* medical practice; medical practitioners
"hermeticism," 3, 9, 64
Hiebner, Israel, 77–78
Hooker, Richard, 121, 127–28
Horace, 131, 133
Horst, George Conard, 93
humoral medicine
　historiography of, 17
　history and philosophy of, 8, 11, 19–24
　"humoral imbalance." *See* humors, constitution in the body
　obstructions and flows, 16–18, 44
humors, 16–18, 24, 27, 50

constitution in the body, 19, 23–25, 38–48, 56–57
movement of, 22, 34
Hyperius, Andreas Gerhard, 58–59

idolatry, 83–84, 119
images and forms, 63–64, 68–71, 73, 82–85
International Classification of Diseases (ICD), 42

Jackson, Henry, 126, 130
James, Gerence, 15n5, 30, 62n7, 65, 95, 111n183, 130
Jennings, Elizabeth, 1–3, 102n100
Jews, 116–17
John the Baptist, 136

Kabbala and Kabbalistic traditions, 65, 73, 75, 110
Kassell, Lauren, 15n5, 33, 36, 52, 71
knowledge. *See also* theology, authoritative Christian knowledge
  acquisition of, 110, 115–19, 125–27, 142–43. *See also* diagnosis; divination
  based on consent, 54, 58–59, 124–30, 137, 141–42
  based on "hard evidence," 55–56
  cumulative, 57–59, 124–25, 142
  divine, 57–58, 94, 109–10, 115–22, 133–35, 141
  transmission of, 57–58, 66, 109–10, 114, 143

Laud, William, 81
leeches, 77, 102, 103, 104, 106, 118
Lilly, William
  astrological practice, 14–15, 66, 86–87
  his autobiography, 5n27, 60–61, 87–88, 92–95, 113–15, 121–22, 123
Luther, Martin, 126–27, 135
Lysons, Daniel and Samuel, 92–93

MacDonald, Michael, 6–7, 11
Madden, Samuel, 92
magic. *See also* angels, conversing with; sigils
  "demonic" vs. "natural," 63, 80–82
  historiography of, 9–10
  texts, 64–65, 110–11
  traditions and categorization of, 62–66, 81–82
medical astrologers. *See* astrological medicine, practitioners of
medical judgement. *See* diagnosis
medical practice
  historiography of, 7–8, 11–12
  history of, 20
medical practitioners, 3, 7, 20, 63. *See also* astrological medicine, practitioners of

medical problems
  blood, voiding of, 18n24, 41, 96
  "falling sickness," 51, 60, 72n66, 72n68, 75n77, 79, 98
  "French pox," 33–34, 43
  "green sickness," 27, 32n109, 104
  lameness, 80, 98, 118
  menstrual disorder, 16, 18, 19, 24, 27, 52
  mental afflictions, 23, 37n157, 61, 79, 98, 99, 103, 104
  nosebleed, 16, 20
  smallpox, 67, 102, 118
  "the mother," 2, 102
  vaginal discharge, 19, 27
  "wind colic," 79
medicine. *See* astrological medicine; diagnosis; disease; humoral medicine; medical practice; medical practitioners; medical problems
Mirandola, Pico della, 110, 133
Moir, George, 93–94
Myddleton, Sir Thomas, 66, 79n100, 103–4, 105
mythology. *See* pagan literature

Napier, Richard
  activities as a practitioner of astrological medicine and divination
    charms, spells and exorcisms, 61–62
    conversing with angels, 2, 87–92, 94–109, 115–19, 121–22
    geomancy, 5, 11
    medical consultations, conduct of, 2, 27–53
    prayer, 11, 62, 88, 91
    sigils, administration of, 3, 65, 71, 77–80, 96–104 *passim*
  assistants and curates, 5, 45, 47, 91n26, 96n62. *See also* Gerence James
  associates, 4–5, 102, 103, 107, 109. *See also* John Dee; Simon Forman; Henry Jackson; Sir Thomas Myddleton; William Page
  biographical information, 4–5
  continuity of treatment, 51–53, 80
  criticism against, 7, 60–61, 81, 83, 119, 123, 130–32
  education and studies, 5
  family and relatives, 4–5. *See also* Sir James Evington; Sir Thomas Myddleton; Sir Robert Napier; Sir Richard Napier; Thomas Napier
  his library, 6, 65, 111
  legacy and reputation, 6–7, 60–61, 66, 86–94, 100
  medical diaries, 2, 5–7, 14–15, 25–36 *passim*, 45–52 *passim*, 62, 72, 86–91, 95–96, 109
  papers and writings, 6

Napier, Richard *(continued)*
  apology for the use of graven images, 83–85
  correspondence, 5n24, 41, 66, 67, 108–9, 115, 117, 123–26, 130, 135–36, 142–43
  "Defence of Astrology," 49, 54, 57–59, 120n260, 131n53, 133n71, 138
  prayers and invocations, 109, 111
  prophecies, 92n32
  theological, 123–24, 126, 128–35
  philosophical and theological outlook, 15, 57–59, 124–39, 141–42
  rector of Great Linford, 5, 123, 131
  residence, 3–5, 130
Napier, Sir Richard, 4–5, 75n77, 79, 107
Napier, Sir Robert (the elder), 4, 5n22
Napier, Thomas, 5–6, 72
Neoplatonism, 64–65, 68, 110, 120–21, 132–34, 141
nosology. *See* disease, taxonomy of

occult qualities, 67–70, 82, 114
occultism, 9, 12, 94n48, 142. *See also* angels; astrology; divination
"Old Rectory." *See* Richard Napier, residence
Oxford, University of, 5

pagan literature, 130–35. *See also* Horace
pagan traditions, 120, 136
Page, William, 104, 124n12, 125n18
pain
  experience and description of, 36–37
  relief of, 20, 49, 51
  setting and movement of, 21–23, 34–37, 57
Paracelsianism, 3n14, 51
patients
  bewitched, 77, 79, 102–3
  complaints and sensations, 33–37, 40–41
  identification of, 25–27
  returning, 37, 40, 51–53
physical examination, 32, 40–41
planets
  association with humors and body parts, 24–25
  association with metals, 67, 72
  daemons and spirits of, 73, 75
  in the astrological chart, 25, 29, 31, 35, 37, 38–39, 57
  symbols of, 65, 73, 75
pregnancy, 27, 103–4
Prideaux, John, 87, 89, 111–12
primitive church. *See* church fathers
prognosis, 40, 95–99
Protestantism, 82–83, 119–22, 124–30, 135, 136, 140, 142

purgatives, 16n8, 20, 36, 46–48, 50, 52, 77, 80, 104, 106
"Puritans" and "Puritanism," 10–11, 60, 81, 83, 124n8, 132. *See also* English Church

Raymond, Joad, 87, 94
Reformation, Christian. *See also* English Church; Protestantism
  historiography of, 9–10
Regiomontanus astrological system, 29
religion. *See* theology
remedies
  choice of, 46–49
  effect of, 52–53
  evacuating humors. *See* bloodletting; emetics; purgatives
  herbal, 47–51, 67, 73, 77, 81, 97
  magical, 62, 81. *See also* sigils, as medicine
  mineral and metallic drugs, 45, 51, 73, 99
  pain soothers, 49, 51
  prescribed by angels, 99–102
  record of giving, 45–46, 50
  timing considerations, 49–50
Resurrection, Day of, 111, 117
Russell, Margaret ("Countess"), 1–2

Satan. *See* devil
Saunders, Richard, 40, 46–49
Sawyer, Ronald, 7, 11, 33–35
Scaliger, Joseph, 136, 138–39
Scott, Sir Walter, 93
scryer. *See* angels, conversing with, using a human medium
second commandment, 82–84
Seitz, Alexander, 22
sickness. *See* disease
sidereal time, 29
sigils
  as medicine, 3, 60, 77–80. *See also* Richard Napier, activities as a practitioner of astrological medicine and divination; sigils, administration of
  fumigation of, 64, 72, 76, 115
  history and philosophy of, 66–70, 81–82
  lawfulness of, 80–85
  making of, 66–67, 71–76
  types of
    constellated rings, 60, 73–75, 79
    Jovian, 3, 72–80 *passim*, 97, 99, 100
    laminas, 72, 73–75
  "sigillated medicine," 73, 79–80
  wrapping of, 60, 77, 79–80
silk. *See* sigils, wrapping of
Skelhorn, Sarah, 87, 114

Speculum Astronomiae, 76, 81, 137
spirits, conjuration of, 64, 87–89, 111, 113, 118–19. *See also* angels, conversing with
Stoics, 133
Storch, Dr. Johannes, 17, 19n30, 20
superstitious belief, 80–83, 88, 92–94
suspicions of witchery. *See* witchcraft accusations.
sweating, 20, 46, 48
sympathy and antipathy, 3, 49, 67, 71, 77, 79

table of houses, 25, 29
talismans. *See* sigils
Taylor, John, 124n16
Temple, Sir Thomas, 41n176, 102
Tertullian, 134
theology. *See also* church councils; church fathers; English Church; Richard Napier, philosophical and theological outlook; Reformation, Christian

and angels, 119–21
authoritative Christian knowledge, 124–30
salvation, 116, 121, 126, 127
Thomas, Keith, 9, 54, 55, 82, 105–6
Tobit, book of, 96, 96n59
Twisse, William, 5n26, 81, 131n56

urine. *See* body, excretions
urine inspection, 32–33, 40, 99

vapors, 17, 22, 34

Wentworth, Lord Thomas, 5, 103, 106, 119
Whitgift, John, 116n226, 128
Williams, John, 81
witchcraft accusations, 1–2, 35, 98n71, 103
Wolphius, Henricus, 136
Wood, Anthony à, 86–87

Yates, Frances, 9

www.ingramcontent.com/pod-product-compliance
Lightning Source LLC
Chambersburg PA
CBHW021945290426
44108CB00012B/967